Title of Series

Biology and Chemistry of Beta Glucan

Volume 01

Beta Glucans - Mechanisms of Action

Editors

Vaclav Vetvicka

University of Louisville
USA

&

Miroslav Novak

Institute of Chemical Technology
Czech Republic

CONTENTS

FOREWORD

Fungi and yeasts represent an important natural source of various biologically active compounds, including polysaccharides. Chitins, α- a β-glucans as well as mannans, xylans and galactans are the principal components of fungal cell walls. Among these polysaccharides, some β-glucans demonstrate significant biological activities. These are mainly branched polysaccharides with a backbone made of β-(1→3) linked glucose sequence and side chains linked by β-(1→6) or β-(1→4) bonds. Solubility in water, as well as other physical and biological features of these polysaccharides, depend on their molecular mass, branching degree, and tertiary structure.

Isolation of β-glucans from raw materials (fungal cell walls) frequently requires special intervention to remove other compounds, *i.e.*, proteins, lipids, polyphenols and other polysaccharides, including various extraction/treatment procedures and preparative chromatography. In this context, control of the purity and structural identification of β-glucans are necessary for the quality control of preparations containing these polysaccharides. Modern spectroscopic and separation methods are effective towards these ends.

β-Glucans isolated from various sources—including yeast, fungi and oat—are known as ballast polysaccharides, which are parts of soluble and insoluble dietary fiber. They are indigestible in the small intestine and can be slightly cleaved by some enzymes of human colon microflora. As dietary fiber components, these polysacchartides show some nutritional benefits such as restriction of indigestion problems, positive influence of cholesterol, fat and glucose metabolism. These properties are instrumental in the preparation of health-improving food supplements containing β-glucans. At present, these polysaccharides are mainly known for their antitumor and immunomodulating activities. Moreover, immunity activation by β-glucans increases the resistance of organisms against viral and microbial infections. The immunomodulating effects of these polysaccharides are important as mild and non-aggressive prevention of tumor metastases and in supporting treatment during the course of chemotherapy. Chemical modifications, *i.e.*, carboxymethylation, phosphorylation or sulphation, are commonly used to improve the water solubility and other features of β-glucans.

This book is devoted to various aspects of β-glucans—composition, structure, and isolation approaches—as well as to their numerous biological activities and description of possible mechanisms of action. For polysaccharide specialists, and also for interested people from other scientific fields, the book brings a broad range of up-to-date information together with outlines of possible future prospects.

Andriy Synytsya
Department of Carbohydrate Chemistry and Technology
Institute of Chemical Technology
Prague 6
Czech Republic

PREFACE

Natural products, useful in preventing and/or treating various diseases, have been sought after throughout the history of humankind. Usually, the mechanisms of the actions remain unknown and the interest of both the public and professionals slowly evaporates. Even β(1-3),(1-6)-D-glucan (with a variable degree of branching; hereafter β-glucan), arguably the most studied natural immunomodulator, did not escape the significant problems. One main problem in characterizing natural products also occurred with β-glucan: these substances usually represent a complex mixture of ingredients, each of which might contribute to the biological activity. The considerable heterogeneity of all natural β-glucans, not only from saccharomycetes but also from other sources, obviously was and continues to be the cause of a series of mutually contradicting conclusions. The heterogeneity is in fact of double character - first, β-glucan from different producing organisms can differ at least in solubility, degree of polymerization and degree of branching, and second, even in the defined organism several types of β-glucan can be present differing in the before mentioned parameters. Therefore, the proper and scientifically well-founded evaluation of β-glucan properties had to focus not only on biochemical characteristics and biological activities of certain isolated preparation but, first of all, on adequate isolation techniques which, at the end, gave us the purified material consisting of a reproducible specimen of pure β-glucan molecules. As a result, the only meaningful data has come from the experiments based on sufficiently purified and chemically standardized β-glucan. The chapter *Development of Views on β-Glucan Composition and Structure* includes several examples of the fact that insufficiently purified β-glucan, together with improper chemical analytical methods, was a cause of erroneous interpretation of β-glucan structure. The same, of course, should be, and in many cases really was, a reason for misinterpretation of the results of biological activities testing of β-glucan.

Chapter *Biological Actions of β-Glucan* brings the newest information about the various mechanisms of interaction of β-glucan with individual cells. This information is supported by a detailed information regarding the specific β-glucan receptors described in Chapter *β-Glucan Receptors*. In Chapter *β-Glucan-mediated Tumor Immunotherapy – Mechanism of Action and Perspectives,* we provide an up-to-date information about the role of β-glucan in tumor therapy—including the summary on current clinical trials. The next chapter *Use of β-glucan for Drug Delivery Applications* reveals a detailed information about the new and extremely promising use of β-glucan in medicine.

β-Glucan is generally considered to be a very safe immunomodulator. However, data shown in Chapter *Detrimental Effects of β(1-3),(1-6)-D-Glucan* suggest potential problems in the case of using β-glucan together with some anti-inflammatory drugs.

In general, it is clear that during the decades of research, numerous types of β-glucan from many natural sources have been isolated and described. Unfortunately, not all of these glucans were created equal and β-glucans widely differ not only in physicochemical properties, such as branching or molecular weight, but also in biological properties. On top of that, some of β-glucans described in literature have no biological activities at all.

The original studies of effects β-glucan done on the immune system, focused on mice. Subsequent studies demonstrated that β-glucan possesses a strong immunostimulating activity in a wide variety of other species, as well including earthworms, shrimp, fish, tortoises, chicken, rats, rabbits, guinea pigs, sheep, pigs, cattle, and, last but not least, humans. Based on these results, it has been concluded that β-glucan represents a type of immunostimulant that is active over the broadest spectrum of biological species and that it is one of the first immunostimulants actively spanning an evolutionary spectrum. Some experiments also show that β-glucan can help even in the protection of plants. β-Glucan is therefore not only a biologically active polysaccharide with strong immunomodulating effects, but is also considered to be an evolutionary and a very old stimulant of a variety of defense immune reactions.

Numerous routes of administration of β-glucan have been tested. These include, *i.a.*, intraperitoneal, intramuscular, sub-cutaneous and intravenous applications. For many years, oral treatment with β-glucan has been on the periphery of interest, despite the fact that it represents the most convenient route. However, in the last decade a renewed interest in human application has brought about some important studies of orally-administered β-glucan that revealed intravenous administration to be dangerous and intramuscular administration to be painful.

Despite of long-term interest and research, the mechanism of how β-glucan affected our health remained in many ways a mystery. Only in the last decade, an extensive research by numerous scientific groups has helped to reveal the extraordinary effects that β-glucan exerts on various physiological and pathophysiological processes of our body. Based on more than 10,000 studies about various types of β-glucan, we can conclude that β-glucans from fungi (both macro- and micromycetes, particularly yeast), and seaweed are well-known biologic response modifiers that function as immunostimulants against infectious diseases and cancer. Unlike most of the other natural products, properly purified β-glucans retain their bioactivity in spite of rather drastic isolation procedures. This allows us to characterize how β-glucan works on a cellular and molecular level.

In the present book, we tried to select and bring together information on both theoretical and practical aspects of β-glucan use. The book is for every reader seeking a prompt and precise information on β-glucan history and various biological effects including possible use in the treatment of various diseases, without having any preliminary knowledge of the field.

In conclusion, we would like to express one wish - that this book not to be regarded as a startling experiment to join the highly remote bioscientific disciplines, such as carbohydrate chemistry, clinical practice and immunology. We presume that in the next few years the fields of natural immunomodulators, and β-glucan in particular, will provide not only an inspiration for further investigation in biomedicine, but also a number of practical applicable results. It is an objective of all the authors of this book to trace the progress in this field and to deliver codified results in a subsequent edition.

Vaclav Vetvicka
University of Louisville

Miroslav Novak
Institute of Chemical Technology

CONTRIBUTORS

M. Novak

Institute of Chemical Technology, Department of Carbohydrate Chemistry and Technology, Prague, Czech Republic

V. Vetvicka

University of Louisville, Department of Pathology, MDR Bldg., Louisville, KY, USA

Alexandra E. Clark

Section of Immunology and Infection, Institute of Medical Sciences, University of Aberdeen, Aberdeen, UK

Ann M. Kerrigan

Section of Immunology and Infection, Institute of Medical Sciences, University of Aberdeen, Aberdeen, UK

Gordon D. Brown

Section of Infection and Immunity, Institute of Medical Sciences, School of Medicine and Dentistry, University of Aberdeen, Aberdeen, AB25 2ZD, UK.

J. Yan

Tumor Immunobiology Program, James Graham Brown Cancer, Department of Medicine, University of Louisville, Louisville, KY, USA

E. Soto

Program in Molecular Medicine, University of Massachusetts Medical School, Worcester, MA, USA

G. Ostroff

Program in Molecular Medicine, University of Massachusetts Medical School, 373 Plantation Street, Worcester, MA 01604, USA

Naohito Ohno

Laboratory for Immunopharmacology of Microbial Products, School of Pharmacy, Tokyo University of Pharmacy & Life Science, 1432-1 Horinouchi, Hachioji, Tokyo 192-0392, Japan

CHAPTER 1

Development of Views on β-Glucan Composition and Structure

Miroslav Novak[1] and Vaclav Vetvicka[2],*

[1]*Institute of Chemical Technology, Prague, Czech Republic and* [2]*University of Louisville, Department of Pathology, Louisville, KY, USA*

Abstract: β-Glucans are well-established and powerful immunomodulators. Their history, however, has been complicated and is full of dead ends. Starting from their chemical composition and structure, continuing with their biological effects and activities, and ending with the mechanism of their function, β-glucans are extensively studied for over 70 years. This Chapter is focused on the history of β-glucans, going back to the half of the 19th century, when β-glucan was believed to be a type of cellulose, and to the 40s of the last century when Shear and Turner described a polysaccharide substance, isolated from *Serratia marcescens*, that caused necrosis of tumors.

INTRODUCTION

For many centuries, fever has been regarded as a mighty and powerful mechanism for fighting all types of disease, including cancer [1]. In the medieval ages, sudatory baths, together with mercurial preparations, were used to cure certain skin diseases, as well as syphilis [2]. The idea of a curative effect by increased temperature has recently led to attempts to use artificial methods to induce fever and the subsequent healing processes. At the beginning of the 18th century, it was already known that certain infectious diseases showed a therapeutic effect on malignant processes. It was believed that it is brought about by elevated temperature. At present, it is known that infections are potent immunomodulators by causing significant alterations in one or more mediators of homeostasis and that these effects are not necessarily accompanied with fever.

Purposeful use of such therapy dates approximately from the middle of the 19th century when Busch [3] used an iatrogenic infection with erysipelas to cure a patient with a soft tissue sarcoma of the neck in the hope that it would cause a tumor regression. This method of therapy was clearly in its pioneering stage. After infection, certain regression of the tumor occurred. However, after a certain time, the tumor again proliferated. Three decades later, in 1881, β-hemolytic streptococci of the group A were identified as the causative agent of erysipelas [4, 5].

Systematic studies of such therapeutical treatment of malignant tumors began in 1891 by William B. Coley [6]. It is likely that he was not aware of Bush's experiment with erysipelas. At first, Coley speculated that the infection around the tumor site induced a direct cytotoxic reaction and in May 1891 he conducted the first treatment of an inoperable tumor with local injection of streptococcal culture [5, 7]. Later, because of the danger of live streptococcal organisms, Coley continued his treatment using less dangerous heat inactivated microbial culture extracts. On the basis of other observations that the presence of *Bacillus prodigiosus* (now renamed to *Serratia marcescens*) can enhance the virulence of streptococci, Coley combined *Streptococcus pyogenes* extract together with *Bacillus prodigiosus* one [8, 9]. This mixture is known as "Coley's toxin" or "Coley's vaccine" [10], which is rarely, if ever, used nowadays. In spite of the fact, that Coley carried out his experiments over a century ago, the mechanism of the toxin remain unexplained until now. It was believed that one of the biologically active agents in Coley's toxin was lipopolysaccharide (LPS) which causes fever leading to an increase of lymphocyte activity and boosting tumor necrosis factor (TNF). This idea is incomplete: LPS, as typical part of an outer membrane of Gram-negative bacteria should originate only from *Serratia*, while streptococci are typically Gram-positive microbes and do not contain LPS; so the presence of LPS cannot explain the effect of a streptococcal component. Tsung and Norton [11] reported more probable findings that efficacious component of Coley's toxin was interleukin-12, rather than LPS and/or TNF.

Another alternative to intact microbes or more or less defined extracts had been used in past. Some better characterized

Address correspondence to: Dr. Vaclav Vetvicka: University of Louisville, Department of Pathology, MDR Bldg., Louisville, KY, USA; Tel: 502-852-1612; E-mail: vaclav.vetvicka@louisville.edu

components of microorganisms caused ferocious reactions in higher organisms and these reactions were comparable to pathophysiological conditions during infection by intact microbes. It is likely that the first investigated substance with these properties was lipopolysaccharide, presented in an outer membrane of Gram-negative bacteria (*e.g.,* of *Escherichia, Salmonella, Serratia, Shigella, Pseudomonas, Neisseria, Haemophilus* genera, and also of phytopathogenic bacteria such as *Agrobacterium* or *Rhizobium spp.*). Perhaps the first paper describing the LPS was published by Billroth in 1865 [12]. Bacterial lipopolysaccharides typically consist of a hydrophobic domain known as lipid A (or endotoxin), a nonrepeating "core" oligosaccharide and a distal polysaccharide (or O-antigen). The core domain always contains an oligosaccharide component which attaches directly to lipid A and commonly contains sugars such as glucose or mannose as well as typical sugar derivatives L-glycero-D-manno-D-heptose and 3-deoxy-D-manno-2-octulosonic acid (also known as 2-keto-3-deoxyoctonate, Kdo) [13] (see Fig. **1**). The O-antigen is attached to the core oligosaccharide; it comprises the outermost domain of the LPS molecule and is exposed on the outer surface of the bacterial cell. The composition of the O-polysaccharide chain varies from strain to strain. More than 60 monosaccharides and 30 different noncarbohydrate components have been recognized [14]. Application of LPS led to increased phagocytosis with a potential protective effect for a host, in both humans and experimental animals. However, its toxic effects (fever, diarrhea, hypotensive shock, intravascular coagulation, multiple organ dysfunctions, *etc.*) completely predominated and impeded a simple use of that substance in medicine. Afterwards it was found that a toxic principle of LPS was its lipid moiety (lipid A), which is a glycolipid typically composed of a 2-amino-2-deoxy-D-glucose (glucosamine) disaccharide backbone. The backbone is phosphorylated and substituting fatty acids can either be linked to the phosphates or can replace them [15]. In comparison with lipid A, the saccharide moieties of LPS are non-toxic and at the same time also bear immunomodulating activity [16]. It was apparent that even polysaccharides themselves could act as immunomodulators, while their toxicity was negligible.

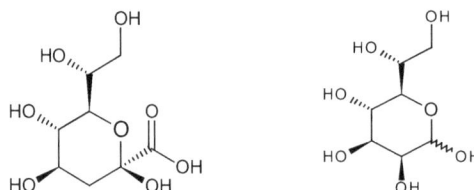

Figure 1: Typical saccharides of the core of LPS (L-glycero-D-manno-D-heptose and 3-deoxy-D-manno-2-octulosonic acid, Kdo)

POLYSACCHARIDES AS IMMUNOMODULATORS

The mentioned finding was probably the catalyst for research of other polysaccharide preparations. Proper history of polysaccharides as immunomodulators goes back to the 40s of the last century when Shear and Turner [17] described a polysaccharide substance, again from *Serratia marcescens* cultures, that caused necrosis of tumors. The first attempt to solve Shear's polysaccharide structure was done by Rathgeb and Sylvén [18]. They fractionated the polysaccharide with trichloroacetic acid and ethanol followed by treatment with hot picric acid. On the basis of methylation analysis and periodate oxidation, they suggested that the polysaccharide was composed of glucose residues combined mutually by alternating 1,4- and 1,6-glycosidic bonds. Results of this work were criticized by Srivastava and Adams [19], since the identification of the methylated sugar fragments was based solely upon paper chromatographic analysis, when methyl ethers of different sugars can have the same R_f value in a particular solvent having the same degree of methylation. Srivastava and Adams subsequently found Shear's polysaccharide to be a mixture of three polysaccharides with the main chain consisting of D-glucose and D-mannose units connected by (1→3) glycosidic linkages. Some glucose residues carried branches at C-2 and C-4, which terminate in either D-glucose or D-mannuronic acid residues. These three polysaccharides had similar chemical structures but varied in the component sugar ratios and degree of branching.

Generally speaking, immunomodulatory preparations from bacteria - either extracts from intact cells or isolated products such as Shear's polysaccharide - are *à priori* suspicious and their use can be dangerous. Perhaps due to this fact, focus was given to polysaccharides isolated from "human friendly" organisms, *i.e.*, yeasts and edible mushrooms. At the same time, Pillemer and Ecker [20] isolated an insoluble fraction from fresh baker's yeast (*Saccharomyces cerevisiae*) which can specifically inactivate the third component of complement (C3); this preparation was named zymin, Ecker's fraction, and lately zymosan.

Pillemer and his coworkers isolated from the human as well animal blood serum a protein that destroys bacteria and neutralizes viruses, which they named properdin (from the Latin term *perdere*, to destroy) [21], and described the relationship of the properdin system to zymosan. This led to studies on the possible effects of zymosan on properdin levels in serum and on host resistance [22, 23, 24, 25, 26]. Nelson [27] originally disputed these findings in 1958, but further studies led to a confirmation of much of Pillemer's work [28]. However, the current knowledge suggests a different conclusion. Despite the fact that some results indicate that zymosan influences host resistance to some extent (for example it activates neutrophils and induces the generation of the interleukin IL-8 [29]), and/or it has been used as an agent for the regression of tumor transplants [30, 31], these effects are probably not mediated through properdin [32]. During these studies Pillemer *et al.* reached the conclusion that the capacity of zymosan to inactivate complement and to stimulate host resistance is not a unique property, since some other polysaccharides behave similarly [33].

Although it was known that zymosan is able to stimulate non-specific immune response, at the onset it was not clear what component of this rather crude composition is responsible for that activity. According to chemical analyses, zymosan contains on average 55 % of glucan, together with about 19% of mannan, 15% of protein, 6% of fat and 3% of ash (*e.g.,* [34]). Pillemer and his coworkers [33] demonstrated that the essential activity of zymosan consists in a glucan-rich fraction. In contrast, the mannan-rich fraction - glucomannan protein - showed little activity. Similar results were obtained by Riggi and Di Luzio [35] who tested the functional activity of the reticuloendothelial system, detected by the intravascular clearance of colloidal carbon, and the degree of induced reticuloendothelial hyperplasia after the intravenous injection of various constituents of zymosan. The mannan component of the yeast cell was inactive and the activity was still present in the zymosan residue after removal of free and bound lipids. The administration of glucan, derived from either whole yeast or its cell wall, resulted in marked reticuloendothelial activation and induced hyperplasia, demonstrating it to be the active reticuloendothelial stimulating agent.

In the experiments mentioned above, identity of "glucan", *i.e.,* polymer of glucose, was detected only by its hydrolysis and subsequent analyses of monosaccharide composition, without any closer exploration of its structure or molecular weight. More detailed insights to structural aspects were solved simultaneously and, in many cases, independently by several researchers or research teams who were focused more either to medicine or to chemistry. Nicholas R. DiLuzio and his coworkers from the Tulane University in New Orleans pioneered further medical research of β-glucan. In a number of papers [36, 37, 38, 39, 40], they demonstrated that glucan administration caused significant phagocytic stimulation of the reticuloendothelial system, enhanced host defense mechanisms, and resistance to experimental tumors. For the sake of accuracy, it should be noted that DiLuzio's group was not the only one that described similar results (*e.g.* [41]).

Intensive research of immuno-modulating activities of β-glucan was also conducted in Japan and they arrived at β-glucan *via* a different route. In Asian medicine, consuming different medicinal mushrooms (shiitake, maitake, reishi, *etc.*) has been a long tradition. Detailed studies of the biological effects of these mushrooms, especially their anticancer action, show β-glucans as a main cause of non-specific immunomodulation. The onset of this investigation is attributed to Goro Chihara from the Teikyo University in Kawasaki [42, 43, 44, 45], who isolated β-glucan, named by him lentinan, from mushroom shiitake (*Lentinus edodes*, now *Lentinula edodes*).

FUNGAL β-GLUCAN

Although β-glucans can be isolated from a number of different natural sources, such as algae, bacteria, cereals, *etc.*, fungi are the most important resource. Polysaccharides in the cell wall of different fungi are in such a degree important constituents that a content of typical wall polysaccharides represents one of the taxonomic parameters (examples in Table **1**). It is important to note, however, that taxonomic conception of kingdom *Fungi* went through significant progression and many microorganisms, earlier classified as *Fungi,* are now categorized in other kingdoms. As shown on this table, it is evident that the most important producers of β-glucans are ascomycetes (where yeasts and certain filamentous moulds pertain) and basidiomycetes (to this class belongs among others most of macromycetes, thus mushrooms - edible or non-edible - whether found in nature or artificially cultivated).

Table 1: Polysaccharide composition of cell walls of different fungal organisms

Kingdom	Class	(Example)		Prevailing polysaccharides in cell walls of vegetative cells
		Order	Genus and species	
Protista	*Acrasiomycetes*		*Dictyostelium discoideum*	cellulose - glycogen
Chromista	*Oomycetes*	*Perenosporales*	*Plasmopara viticola* *Phytophthora infestans*	cellulose - **β-glucan**
	Hyphochytridio- mycetes	*Hyphochytriales*	*Rhizidiomyces parasiticus*	cellulose - chitin
Fungi	*Zygomycetes*	*Mucorales*	*Mucor mucedo*	chitosan - chitin
	Chytridiomycetes	*Chytridiales*	*Blastocladiella emersonii*	
	Ascomycetes	*Eurotiales*	*Aspergillus niger*	chitin - **β-glucan**
	Homobasidio- mycetes	*Agaricales*	*Agaricus bisporus* *Lentinula edodes* *Pleurotus ostreatus*	
		Stereales	*Grifola frondosa* *Schizophyllum commune*	
	Hemiascomycetes	*Saccharomycetales*	*Saccharomyces cerevisiae* *Schizosaccharomyces octosporus*	mannan - **β-glucan**
	Hypomycetes	*Sporobolo- mycetaceae*	*Sporobolomyces roseus*	mannan - chitin

EXPLORING OF β-GLUCAN COMPOSITION AND STRUCTURE

In all probability, the first attempt to isolate β-glucan from yeast was done by Salkowski [46] by boiling pressed yeast with 3 % KOH and treating a residue successively with water, dilute hydrochloric acid, again with water, alcohol, and ethylether, and finally extracting with ethylether in a Soxhlet's apparatus to remove fat. This resulted in a yellowish powder he called "yeast cellulose". The powder could be partially solubilized by heating with water under pressure while remaining an insoluble residue; the solution gave an intense brownish-red color with iodine solution and was by him named erythrocellulose, whereas the residue had not reacting with iodine was termed achroocellulose. Elemental analysis of this "yeast cellulose" point to the formula $C_6H_{10}O_5$, and on treatment with dilute acids glucose was formed. Van Wisselingh held a similar opinion [47] and stated that in fungal cell wall either cellulose or chitin could prevail. Sevag's group [48] confirmed elementary composition of polysaccharide isolated analogically as Salkowski's "yeast cellulose". These authors also considered that composition and quantity of carbohydrates from yeast would differ from strain to strain.

In the first half of the 20th century, only moderate progress was made from the time of Salkowski. Zechmeister and Toth [49, 50] studied the insoluble yeast polysaccharide, described by Salkowski as "yeast cellulose" and found that it did not give the characteristic cellulose tests, *i.e.*, blue color with an iodine solution, it was not soluble in tetraamminediaquacopper dihydroxide (Schweizer's reagent) and after hydrolysis it did not yield cellobiose. As an important result, it can be mentioned that the authors for the first time demonstrated a rather unusual (1-3)-glycosidic linkage between the anhydroglucose units in the polysaccharide chain.

Hassid *et al.* [51] studied the structure of this insoluble yeast polysaccharide and suggested that the glucosidic linkages of the anhydroglucose units are predominantly of the β-type. On methylation analysis of the polysaccharide they obtained 2,4,6-*O*-trimethylglucose as the sole product of hydrolysis, what confirmed presence of (1→3)-glycosidic linkages. Unfortunately, in this paper an erroneous presumption was made that this polysaccharide had the closed chain. The reason for this error was the fact that they did not find any 2,3,4,6-*O*-tetramethylglucose, denoting the nonreducing-end glucose. In addition, although they stated the presence of β-linkage in the glucan, the proposed structure was drawn with α-linkages (Fig. **2**). This was surely an oversight.

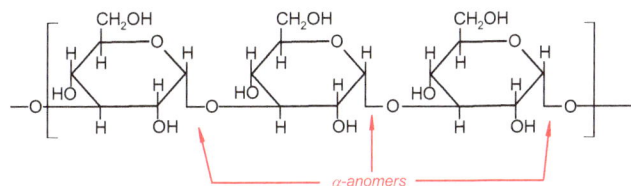

Figure 2: Original scheme of "glucan" structure from Hasid *et al.* [51].

Barry and Dillon [52], by periodate oxidation, again observed that the polysaccharide obtained from yeast by extraction with diluted alkali is a (1→3)-glucan. They found that glucan was only slightly oxidizable by periodate. This confirmed the existence of (1-3)-linkages in the main chain: periodate oxidizes only vicinal hydroxyl groups and, in this case, these are only present in the end glucose molecules (see Fig. **3**). Isolated glucan by hydrolysis with fuming HCl at normal temperature gave oligosaccharides which were hydrolyzed to glucose by emulsin (β-glucosidase - active on β-stereoisomers), but not by taka-diastase (mixture of amylases - active on α-stereoisomers). From these data, it was possible to judge that the isolated polysaccharide had a β-configuration.

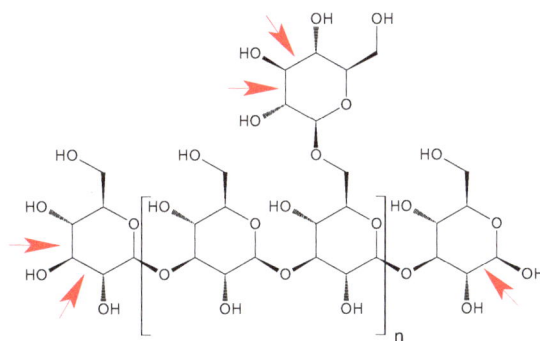

Figure 3: Points of possible periodate attack on β-glucan molecule (marked by arrows).

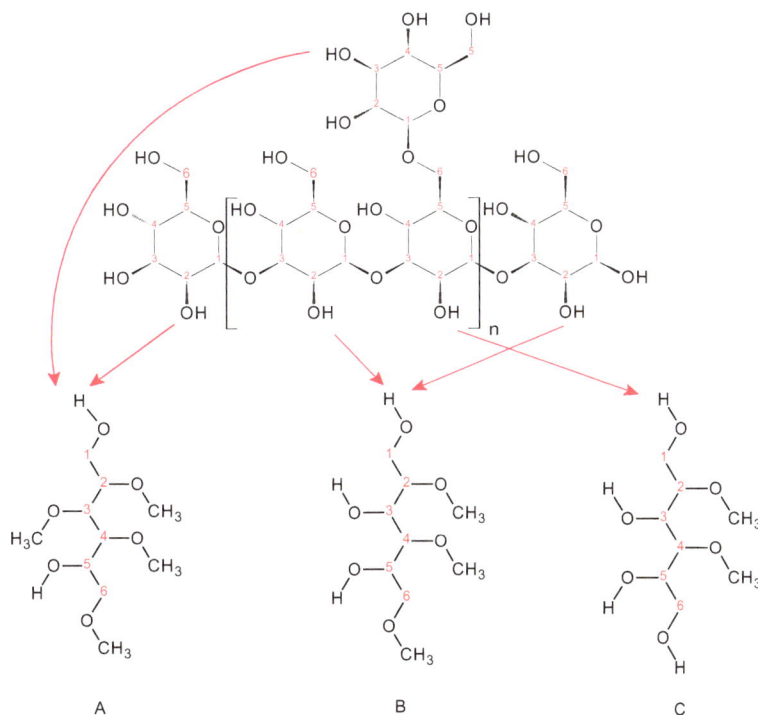

Figure 4: Permethylated *O*-methylglucitols from β-glucan (A - 2,3,4,6-*O*-tetramethyl-, B - 2,4,6-*O*-trimethyl-, C - 2,4-*O*-dimethylglucitol).

Bell and Northcote [53], using methylation analysis, were the first who described yeast β-glucan as branched polysaccharide, however the branching position was identified erroneously. Namely, methylation of β-glucan is difficult and one has to keep in mind that around the half of the last century, a proper method for β-glucan methylation analysis was not yet developed. Due to this fact, some papers brought rather discrepant results. Bell and Northcote found 2,3,4,6-*O*-tetramethyl-, 2,4,6-*O*-trimethyl-, and 4,6-*O*-dimethyl-D-glucose, corresponding to nonreducing-end glucose molecules (2,3,4,6-*O*-tetramethyl-D-glucose), glucoses on reducing-end and in inner parts of chain (2,4,6-*O*-tetramethyl-D-glucose) and glucoses in the branching points (4,6-*O*-dimethyl-D-glucose). But the 4,6-*O*-dimethyl-D-glucose would be originated from a moiety branched by β(1-2)-linkage, which mismatches with reality (for that a 2,4-*O*-dimethyl derivate should be detected - see Fig. **4**). The glycosidic linkage (1-2) mentioned above, which is according temporary knowledge not present in fungal β-glucans, was described by several other authors as well. Houwink *et al.* [54] had this type of linkage detected on the basis of the rentgenographic method and Northcote [55] was convinced on branching (1→2) still more than a decade later.

Contrary to these results, it was concluded later by Peat *et al.* [56, 57] that β-glucan is not branched - the main chain is linear and consists of β-D-glucopyranoses randomly linked by (1-3)- and (1-6)-linkages. They found no evidence to indicate the presence of (1-2)-linkages. To determine (1-6)-linkages, the authors proposed an alternative method according to Oldham and Rutherford [58] - estimating the number of primary hydroxyl groups presented in branches by esterification with toluene-*p*-sulfonyl chloride (tosylation) followed by replacing of tosylated primary hydroxyls by easily analyzable iodine atoms. These results are in sharp contradiction with many other findings on the highly branched character of β-glucan and turned out to be erroneous later, *e.g.*, by Misaki *et al.* [59], who showed apodeictically, on the basis of a number of rigorous methylation and oxidation analyses, that β-glucan contains main chain of β(1-3)-D-glucoses with branches at C-6. Similarly, detailed fractionation of cell wall components and their characterization was made by Manners *et al.* [60]. Most important - and concluding - study leading toward fungal β-glucan structure was probably described in paper of Manners *et al.* [61], who confirmed the significant heterogeneity of different preparations of glucan and possible discrepancies arising from that. The current belief is that the main component of β-glucan from the yeast cell wall is a slightly branched, high-molecular (1-3)-β-D-glucan (DP about 1500, molecular weight ca. 240 kDa), with about 3% of β(1-6) branching. Generally accepted structure of β-glucan is shown in Fig. **5**.

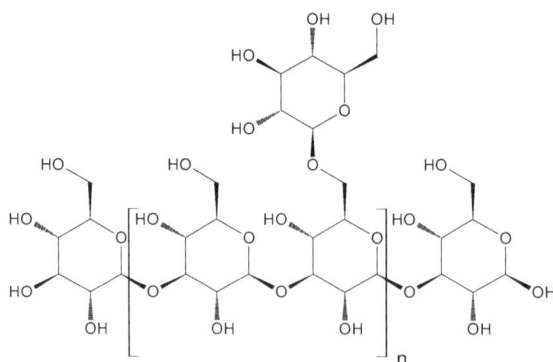

Figure 5: Structure of fungal (yeast) β-glucan. Degree of branching generally depends on genus, species and strain of the organism.

CONCLUSION

It is currently accepted, without doubt, that β-glucans are powerful immunomodulators. However, between the first application of immunotherapeutical principles and contemporary knowledge of β-glucan physiological effects lies a rather confusing route. The same holds true, and perhaps to even a greater extent, regarding its chemical composition and structure. Based upon existing literature, the beginning of immunotherapy can be traced back several centuries. From the onset of fever therapy, when in addition to the germicidal effect of increased temperature the body's defense mechanisms appearing to work more efficiently, therapeutists looked for less destructive methods of treatment. As a promising way for that purpose, knowledge that certain infections led to a curative effect was exploited.

Initially, practiced iatrogenic infections were replaced by the application of more or less defined extracts from microbial cultures of pathogenic bacteria (Coley's toxin). As a possible alternative to rough preparations from bacterial cultures, certain more-defined bacterial products were tested. Undoubtedly, the first preparation used in this manner was lipopolysaccharide from Gram-negative bacteria. In spite of a certain positive effect on the immune system, its toxicity prevented any practical application. Findings that LPS contains an active non-toxic polysaccharide moiety perhaps initiated the research of immunomodulatory polysaccharides. Proper history of polysaccharides as immunomodulators goes back to the 40s of the last century when Shear and Turner described a polysaccharide substance, isolated from *Serratia marcescens*, which caused necrosis of tumors.

Concurrent with the research of bacterial polysaccharides, polysaccharides from more "human friendly" fungi, primarily from yeast or edible and/or medicinal mushrooms, were explored. In a rather crude preparation from intact yeast cells known as zymosan, β-glucan was identified as its most immunologically active component. The same conclusion resulted from research on Asian medicinal mushrooms. During several decades, β-glucan proved to be the most active and promising fungal immunomodulatory polysaccharide, what has resulted in intensive study.

From beginning of the 19th century, when β-glucan was believed to be a type of cellulose, the composition and structure of β-glucan were examined step by step and sometimes erroneously. However, at the study's conclusion, the proper structure was determined. Unfortunately, these final results do not erase β-glucan's major problem - its considerable heterogeneity, preventing precise results when used in medicine or in basic research. This issue could be solved either by very meticulous methods of preparation, or by using semisynthetic and synthetic probes suitable for accurate immunological research. There is also the future possibility of substituting natural β-glucan.

The problems mentioned in the pursuit of the correct composition and structure of β-glucan were caused by many circumstances. One of them, and perhaps the most influential during this period of prospecting, exists in the fact that early methods of sugar analyses were not very precise and/or sufficient for non-soluble β-glucans. For example, it is probable that many formerly published results obtained by the methylation analysis were influenced by incomplete methylation, succeeded by tedious and inaccurate detection of methylated glucitols carried out by paper chromatography and verified by their melting points. Informative values of the results were essentially improved by enhanced chemical procedures and implementation of gas chromatography for methylated glucitol detection.

Non-solubility and heterogeneity of β-glucan represented another serious obstacle. Considerable heterogeneity of all natural β-glucans, not only from saccharomycetes but also from other sources, obviously was and continues to be a cause of a series of mutually contradicting conclusions. Recently, attempts were made to solve this problem using semisynthetic and synthetic probes suitable for accurate immunological research and, in the future, possibly substituting natural β-glucan [62, 63].

ACKNOWLEDGMENT

The work of Miroslav Novak was supported by the grant of the Czech Science Foundation No. 525/09/1033.

REFERENCES

[1]   Hobohm U. Fever and cancer in perspective. Cancer Immunol Immunother 2001; 50:391-6.
[2]   Paracelsus Th. B. Von der Frantzösischen kranckheit drey Bücher. Frankfurt a/M., Hermann Gülferich, 1553.
[3]   Busch W. Verhandlungen artzlicher gesellschaften. Berl Klin. Wochenschr 1868; 5:137-8.
[4]   Senn N. The treatment of malignant tumors by the toxins of the streptococcus of erysipelas. J Amer Med Assoc 1895; 25:131-4.
[5]   Bickels J, Kollender Y, Merinsky O, Meller I. Coley's toxin: Historical perspective. Israel Med Assoc J 2002; 4:471-2.
[6]   Coley WB. Contribution to the knowledge of sarcoma. Ann Surg 1891; 14:199-220.
[7]   Coley WB. The treatment of malignant tumors by repeated inoculations of erysipelas: with a report of ten original cases. Am J Med Sci 1893; 105:487-511.
[8]   Coley WB. Treatment of inoperable malignant tumors with the toxines of erysipelas and the bacillus prodigiosus. Am J Med Sci 1894; 108:50-66.
[9]   Coley WB. The treatment of inoperable sarcoma with the mixed toxins of erysipelas and bacillus prodigiosus - immediate and final results in one hundred and forty cases. J Amer Med Assoc 1898; 31:289-342.
[10]   Wiemann B, Starnes CO. Coley's toxins, tumor necrosis factor and cancer research: a historical perspective. Pharmacol Ther 1994; 64: 529-64.

[11] Tsung K, Norton JA. Lessons from Coley's Toxin. Surg Oncol 2006; 15:25-8.

[12] Billroth T. Beobachtungs-Studien über das Wundfieber und accidentelle Wundkrankheiter. Arch Klin Chir 1865; 6:372-99.

[13] Holst O. The structures of core regions from enterobacterial lipopolysaccharides - an update. FEMS Microbiol Lett 2007; 271:3-11.

[14] Brade H, Opal SM, Vogel SN, Morrison DC, eds., Endotoxin in health and disease. New York/Basel: Marcel Dekker 1999; pp. 155-78.

[15] Raetz CRH, Whitfield C. Lipopolysaccharide endotoxins. Annu Rev Biochem 2002; 71:635-700.

[16] Nowotny A. Molecular aspects of endotoxic reactions. Bact Rev1969; 33:72-98.

[17] Shear MJ, Turner FC. Chemical treatment of tumours. V. Isolation of the hemorrhage-producing fraction from *Serratia marcescens* (Bacillus prodigiosus) culture filtrate. J Natl Cancer Inst 1943; 4: 81-97.

[18] Rathgeb P, Sylvén B. Fractionation studies on the tumor necrotizing agent from *Serratia marcescens* (Shear's polysaccharide). J Natl Cancer Inst 1954; 14:1099-1108.

[19] Srivastava HC, Adams GA. Constitutions of polysaccharides from *Serratia marcescens*. Can J Chem 1962; 40:1415-24.

[20] Pillemer L. Ecker EE. Anticomplementary factor in fresh yeast. J Biol Chem 1941; 137:139-42.

[21] Pillemer L, Blum L, Lepow IH, Ross OA, Todd EW, Wardlaw AC. The properdin system and immunity. I. Demonstration and isolation of a new serum protein, properdin, and its role in immune phenomena. Science 1954; 120:279-85.

[22] Pillemer L, Ross OA. Alterations in properdin levels following injection of zymosan. Science 1955; 121:732-3.

[23] Rowley D. Stimulation of natural immunity to *Escherichia coli* infection: observations on mice. Lancet 1955; 268:232-4.

[24] Landy M, Pillemer L. Elevation of properdin levels in mice following administration of bacterial lipopolysaccharides. J Exp Med 1956; 103:823-33.

[25] Pillemer L. The properdin system. Trans N Y Acad Sci 1955; 17:526-30.

[26] Pillemer L, Blum L, Lepow IH, Wurz L, Todd EW. The properdin system and immunity. III. The zymosan assay of properdin. J Exp Med 1956; 103:1-13.

[27] Nelson R. An alternative mechanism for the properdin system. J Exp Med 1958; 108:515-35.

[28] Kemper C, Hourcade DE. Properdin: new roles in pattern recognition and target clearance. Mol Immunol 2008; 45: 4048-56.

[29] Au B-T, Teixeira MM, Collins PD, Williams TJ. Effect of PDE4 inhibitors on zymosan-induced IL-8 release from human neutrophils: synergism with prostanoids and salbutamol. British J Pharmacol 1998; 123:1260-66.

[30] Mankowski ZT, Yamashita M, Diller LC. Effect of *Candida guilliermondi* polysaccharide on transplantable mouse sarcoma 37. Proc Soc Exp Biol Med 1957; 96:79-80.

[31] Maeda YY, Chihara G, Ishimura K. Unique increase of serum proteins and action of antitumor polysaccharides. Nature 1974; 252: 250-2.

[32] Pontieri GM, Plescia OJ, Nickerson WJ. Inactivation of complement by polysaccharides. J Bacteriol 1963; 86:1121-2.

[33] Pillemer L, Schoenberg MD. Blum L, Wurz L. Properdin system and immunity. II. Interaction of the properdin system with polysaccharides. Science 1955; 122:545-9.

[34] Di Carlo FJ, Fiore JV. On the composition of zymosan. Science 1958; 127:756-7.

[35] Riggi SJ, Di Luzio NR. Identification of a reticuloendothelial stimulating agent in zymosan. Am J Physiol 1961; 200:297-300.

[36] DiLuzio NR, Pisano JC, Saba TM. Evaluation of the mechanism of glucan-induced stimulation of the reticuloendothelial system. J Reticulo-endothel Soc 1970; 7:731-42.

[37] Di Luzio NR, Riggi SJ. Effects of laminarin, sulfated glucan, and oligosaccharides of glucan on reticuloendothelial activity. J Reticulendothel Soc 1970; 8:465-73.

[38] DiLuzio NR, Hoffmann EO, Cook JA, Browder W, Mansell WA. Glucan-induced enhancement in host resistance to experimental tumors. Prog Cancer Res Therap 1977; 2:475-99.

[39] DiLuzio NR. Immunopharmacology of glucan: A broad spectrum enhancer of host defense mechanisms. Trends Pharmacol Sci 1983; 4:344-7.

[40] Williams DL, Al-Tuwaijri A, DiLuzio NR. Influence of glucan on experimental infections in mice. Int J Immunopharmacol 1980; 2:189-98.

[41] Filkins JP, Lubitz JM, Smith JJ. The effect of zymosan and glucan on the reticuloendothelial system and on resistance to traumatic shock. Angiology 1964; 15: 465-70.

[42] Chihara G, Maeda YY, Hamuro J, Sasaki T, Fukuoka F. Inhibition of mouse sarcoma 180 by polysaccharides from *Lentinus edodes* (Berk.) Sing. Nature 1969; 222:687-8.

[43] Chihara G, Maeda YY, Hamuro J. Current status and perspectives of immunomodulators of microbial origin. Int J Tiss Reac 1982; 4:207-25.

[44] Chihara G, Taguchi T. Lentinan: Biological activities and possible clinical use. Riv Immunol Immunofarmacol 1982; 2:93-104.

[45] Chihara G. Immunopharmacology of lentinan and the glucans. Riv Immunol Immunofarmacol 1984; 4:85-96.

[46] Salkowski E. Über die Kohlenhydrate der Hefe. Ber Deutsch Chem Gesell 1894; 27:3325-29.

[47] van Wisselingh C. Mikrochemische Untersuchungen über die Zellwande der Fungi. Jahrb Wiss Botan 1898; 31:619-87.

[48] Sevag MG, Cattaneo C, Maiweg L. Über die Natur der Hefe-Polysaccharide. Justus Liebigs Ann Chem 1935; 519:111-24.

[49] Zechmeister L, Toth G. Über die Polyose der Hefemembran. Biochem Zeitschr 1934; 270:309-16.

[50]   Zechmeister L, Toth G. Über die Polyose der Hefemembran II. Biochem Zeitschr 1936; 284:133-8.

[51]   Hassid WZ, Joslyn MA, McCready RM. The molecular constitution of an insoluble polysaccharide from yeast, *Saccharomyces cerevisiae*. J Amer Chem Soc 1941; 63:295-8.

[52]   Barry VC, Dillon T. On the glucan of the yeast membrane. Proc Royal Irish Acad Sect B, 1943/1944; 49:177-85.

[53]   Bell DJ, Northcote DH. Structure of a cell-wall polysaccharide of baker's yeast. J Chem Soc 1950; 1944-47.

[54]   Houwink AL, Kreger DR, Roelofsen PA. Composition and structure of yeast-cell walls. Nature 1951; 168: 693-4.

[55]   Northcote DH. The structure and organization of the polysaccharides of yeast. Pure Appl Chem 1963; 7:669-75.

[56]   Peat S, Whelan WJ, Edwards TE. Polysaccharides of baker's yeast. Part II. Yeast glucan. J Chem Soc 1958; 3862-8.

[57]   Peat S, Turvey JR, Evans JM. Polysaccharides of baker's yeast. Part III. The presence of 1: 6-linkages in yeast glucan. J Chem Soc 1958; 3868-70.

[58]   Oldham JWH, Rutherford JK. Method for the identification and estimation of the 6-hydroxyl group in glucose. J Amer Chem Soc 1932; 54:366-78.

[59]   Misaki A, Johnson J Jr, Kirkwood S, Scaletti JV, Smith F. Structure of the cell-wall glucan of yeast (*Saccharomyces cerevisiae*). Carbohydr Res 1968; 6:150-64.

[60]   Manners DJ, Masson AJ, Patterson JC, Bjørndal H, Lindberg B. The structure of a β-(1-6)-D-glucan from yeast cell walls. Biochem J 1973; 135:19-36.

[61]   Manners DJ, Masson AJ, Patterson JC. The heterogeneity of glucan preparations from the walls of various yeasts. J Gen Microbiol 1974; 80:411-7.

[62]   Descroix K, Ferrieres V, Jamois E, Yvin JC, Plusquellec D. Recent progress in the field of β-(1,3)-glucans and new applications. Mini-Rev Met Chem 2006; 6: 1341-9.

[63]   Descroix K, Vetvicka V, Laurent I, Jamois F, Yvin JC, Ferrières V. New oligo-β-(1,3)-glucan derivatives as immunostimulating agents. Bioorg Med Chem 2010; 18:348-57.

CHAPTER 2

Biological Actions of β-Glucan

Vaclav Vetvicka[1]* and Miroslav Novak[2]

[1]University of Louisville, Department of Pathology, Louisville, KY, USA and [2]Institute of Chemical Technology, Prague, Czech Republic

Abstract: β-Glucan is a well-known biological response modifier (BRM) that has been used as an adjuvant therapy for cancer since 1980, primarily in Japan. It represents a class of fungal cell wall polysaccharides that are made up entirely of glucose that is β(1-3)-linked together in linear chains with a variable frequency of β(1-6)-linked side chains. β-glucans from cereal grains such as oats or barley can have a similar BRM activity in tumor models but contain linear chains with β(1-4)-linkages in addition to β(1-3)-linkages. β-Glucans also enhance the innate host defense against certain bacteria, yeast, and viral pathogens. In the 1990s, there were attempts in the U.S. to develop a yeast β-glucan as an anti-infective BRM. In addition, β-glucans are also considered to be important in prophylaxis against irradiation. In summary, β-glucan might be the most in important natural immunomodulator.

INTRODUCTION

Natural products, useful in preventing and/or treating various diseases, have been sought after throughout the history of mankind. A major problem in characterizing natural products also occurred with β-glucans, *i.e.,* in nature, they represent a complex mixture of ingredients, each of which might contribute to biological activity. Thirty-six years ago, β-glucans were first described as biological response modifiers (BRM) that could stimulate tumor rejection in mice. As with many other BRM, they were classified as "non-specific" because their molecular target(s) were unknown and their effects appeared to be pleiotropic and unpredictable. Nevertheless, there is extensive literature about the activity of β-glucans in animal tumor models and, for the past 25 years, Japan has used several forms of mushroom-derived β-glucan in cancer patients. The various β-glucans are isolated mostly from yeast, mushrooms and seaweed (some are shown in Table **1**). The sources of β-glucans are strongly based on the traditions and their availability in individual countries. Not surprisingly, the biological activities of β-glucan depend, at least in part, on biochemical characteristics of β-glucan molecule.

Table 1: Some β-Glucans with Immunomodulatory Effects

Name	*Source*	*Character of polymer*
Curdlan	*Alcaligenes faecalis*	*linear*
Laminaran	*Laminaria sp.*	*linear*
Pachymaran	*Poria cocos*	*linear*
Lentinan	*Lentinus edodes*	*branched*
Pleuran	*Pleurotus ostreatus*	*branched*
Schizophyllan	*Schizophyllum commune*	*branched*
Sclerotinan	*Sclerotinia sclerotiorum*	*branched*
Sclero-β-glucan	*Sclerotium glucanicum*	*branched*
Grifolan	*Grifola frondosa*	*branched*
T-4-N	*Dictyophora indusiata*	*branched*
Yeast β-glucan	*Saccharomyces cerevisiae*	*branched*
Barley β-glucan	*Hordeum vulgare*	*branched*

*****Address correspondence to: Dr. Vaclav Vetvicka,** University of Louisville, Department of Pathology, MDR Bldg., Louisville, KY, USA; Tel: 502-852-1612; E-mail: vaclav.vetvicka@louisville.edu

Mechanisms of β-Glucan Action

Despite long-term interest and research, the mechanism of how β-glucan affects our health remained a mystery for a long time. Only in the last decade, extensive research by numerous scientific groups has helped to reveal the extraordinary effects that β-glucan has on our immune system. Experiments done by a University of Louisville research group, led by Gordon Ross, focused on one particular receptor called complement receptor type 3 (CR3 receptor) as a promising target of β-glucan.

This group's research has shown that CR3 serves as a major receptor for β-glucans with human [1] or mouse [2] leukocytes, and is probably responsible for numerous functions of β-glucans. Unlike other non-specific modifiers, β-glucan specifically targets macrophages, neutrophils, and NK cells to tumors that are opsonized with antibodies and C3, and therefore β-glucan has the same specificity as the tumor-opsonizing antibodies. Further research has shown soluble β-glucans that bind to CR3 monogamously and prime the receptor for subsequent cytotoxic activation if, and only if, membrane CR3 are subsequently clustered by contact with the iC3b coating a tumor cell.

CR3, known also as Mac-1, $\alpha_M \beta_2$-integrin, or CD11b/CD18, is a type one membrane glycoprotein made up of two non-covalently linked α and β subunits known as CD11b or α_M (165 kD) and CD18 or β_2 (95 kD). CR3/Mac-1 has two major functions: as the Mac-1 it mediates the diapedesis of phagocytes and NK cells into sites of inflammation by generating a high-affinity binding site for the intercellular adhesion molecule-1 (ICAM-1; CD54) expressed by stimulated endothelium [3]. As CR3, it triggers phagocytosis and degranulation in response to microorganisms or immune complexes opsonized with the complement protein iC3b [4, 5].

The recognition of iC3b on microorganisms by the CR3 of phagocytes and NK cells triggers phagocytosis and cytotoxic degranulation responses that are important in host defense. Stimulation of cytotoxic degranulation by iC3b-opsonized cells requires the presence of specific microbial polysaccharides that are recognized by a lectin domain contained within CR3 that maps to a region that is distinct from its iC3b-binding site. This mode of action represents a form of innate pattern recognition that allows discrimination between microorganisms and host cells. A key finding was that β–glucans could bind to the lectin domain of CR3 and prime the receptor for cytotoxic degranulation in response to tumors that bore iC3b and were normally resistant to this form of cellular cytotoxicity. Many human tumors generate an immune response that results in the deposition of antibody and iC3b on membrane surfaces. This iC3b on tumors serves as a specific target for CR3-bearing neutrophils, macrophages, eosinophils, and NK cells that has been primed with soluble β-glucan. When tumors lack such iC3b, research with mouse tumor models has demonstrated that monoclonal antibodies to tumor antigens can be administered in combination with β-glucan to restore tumor-bound iC3b and assure tumor-specific targeting. Normal tissues surrounding the tumor cells are spared from leukocyte attack because they lack this targeting iC3b.

More than 20 years ago, it was shown that CR3-dependent phagocytosis or degranulation in response to iC3b-opsonized yeast required ligation of two distinct sites in CR3—one for iC3b and another site for β-glucans [6]. Later research mapped these two sites to CD11b. All protein ligands bind to overlapping sites contained within the I-domain of CD11b [7, 8]. The C-terminal region was found to contain the lectin-like site for β-glucans. First, using flow cytometry with FITC-labeled β-glucan and CHO cells expressing recombinant chimeras between CD11b and CD11c, the lectin site was mapped to a region of CD11b located C-terminal to the I-domain [2]. Next, using baculovirus, rCD11b fragments from which the I-domain had been deleted were shown to have the same affinity for ^{125}I-β-glucan when expressed on insect cells without CD18 as the neutrophil CR3 heterodimer did. The lectin domain of CD11b appears to be critically responsible for regulating both the adhesion and cytotoxic degranulation responses of CR3.

The C-terminal location of the lectin site was confirmed in a study of rCR3 binding to *C. albicans* that suggested that ligation of candida polysaccharides to the lectin site caused an increased affinity of protein-binding sites in the I-domain [9]. These studies implicated a site located C-terminal to both the I-domain and the divalent-cation binding repeats that became covered when monoclonal antibodies were attached to the distal I-domain [2,10]. Further studies with rCD11b expressed on insect cells showed that the binding of β-glucan-FITC or ^{125}I-β-glucan could be blocked by antibodies to the I-domain, as well as by monoclonal antibodies to C-terminal epitopes [11].

Binding of β-glucan to a receptor activates macrophages. The activation consists of several interconnected processes including increased chemokinesis, chemotaxis, migration of macrophages to particles to be phagocyted, degranulation leading to increased expression of adhesive molecules on the macrophage surface, adhesion to the endothelium, and migration of macrophages to tissues. In addition, β-glucan binding also triggers intracellular processes, characterized by the respiratory burst after phagocytosis of invading cells (formation of reactive oxygen species and free radicals - hydrogen peroxide, super oxide radical, NO, HClO, HIO, *etc.*), increasing of content and activity of hydrolytic and metabolic enzymes, and signaling processes leading to activation of other phagocytes and secretion of cytokines and other substances initiating inflammation reactions (*e.g.*, interleukins IL-1, IL-9, TNF-α). For an excellent review regarding interaction of β-glucans with macrophages see [12].

For a long time, there was a controversy regarding the way of application, with either intraperitoneal or intravenous application being considered adequate. Subsequent experiments showed that orally-given β-glucan is active similar to an injected one [13]. Some detailed studies have revealed that after administering oral β-glucan (either soluble barley β-glucan or particulate yeast β-glucan) for three days, labeled β-glucan within splenic macrophages appeared to be the same size as the starting material. With time, β-glucan aggregates were found to be concentrated at the edges of the cytoplasm near the cell membrane. Subsequent studies using cultured macrophage cell line examined the fate of labeled β-glucan added to the cells. These experiments showed that phagocytosed β-glucan was slowly degraded within cells and that soluble biologically highly-active fragments of β-glucan were released into the surrounding cells. Complete macrophage degradation of β-glucan required approximately thirteen days. Usually, β-glucan particles remained intact for four days and appeared to fragment into smaller parts and soluble material during ensuing days [14]. Similar results were later independently reached by Chan's group [15].

Effects of β-Glucans on Infection

For a long time, β-glucans have been studied in infections. Using several experimental models, it has been well established that β-glucan protects against infection with both bacteria and protozoa and enhances antibiotic efficacy in infections with antibiotic-resistant bacteria. The protective effect of β-glucans was shown in experimental infection with *Leishmania major* [16] and *L. donovani* [17], *C. albicans* [18], *Toxoplasma gondii* [19], *Streptococcus suis* [20], *Plasmodium berghei* [21], *Staphylococcus aureus* [22], *Escherichia coli* [23], *Mesocestoides corti* [24], *Trypanosoma cruzi* [25], *Eimeria vermiformis* [26] and anthrax infection [27].

In addition to hundreds of reports on β-glucans stimulating the immune response against many infections, there have been numerous studies, including clinical trials, conducted with β-glucan and infections in humans. In the 1990s, Alpha-Beta Technologies conducted a series of human trials. Using double blind, placebo-controlled trials, these studies showed that patients who received β-glucan had significantly less infections, had a decrease in the use of antibiotics, and a shorter stay in the intensive care unit [28].

Browder *et al.* [29] described stimulation of human macrophages in trauma patients and found that β-glucan therapy strongly decreased septic morbidity. A multicenter, double blind study found the optimal dosage of β-glucan in high-risk surgical patients. In addition, these studies demonstrated the safety and efficacy of β-glucan in surgical patients who underwent major thoracic or abdominal surgery. Since no adverse drug experiences associated with β-glucan infusion have been found, β-glucan-treated patients had significantly lower levels of infections. The biological effects of β-glucan on anti-infectious immunity are two-fold: macrophages are activated to produce substances (such as H_2O_2) directly killing the bacteria and stimulation of B lymphocytes to produce more antibodies.

Sepsis leads to the damage and dysfunction of various organs. One of the underlying mechanisms is thought to be the oxidative damage due to the generation of free radicals. Sener's group investigated the putative protective role of yeast-derived β-glucan against sepsis-induced oxidative organ damage. Sepsis was induced by caecal ligation and puncture in rats. Sham operated and sepsis groups received saline or β-glucan once daily for 10 days and 30 minutes prior to and 6 hours after the puncture. Sixteen hours after the surgery, the rats were decapitated and the biochemical changes were determined in the brain, kidney, heart, liver and lung tissues. Tissues were also examined under a light microscope to evaluate the degree of sepsis-induced damage. The results demonstrate that sepsis significantly decreased GSH levels and increased the MDA levels and MPO activity causing oxidative damage. Elevated plasma TNF-alpha levels in septic rats significantly reduced to control levels in the β-glucan-treated rats. Since β-glucan administration reversed these oxidant responses, it seems likely that β-glucan protects against sepsis-induced oxidative organ injury [30]. Similar conclusions were reached later on an independent model [31].

Effects of β-Glucan on Cancer

Based on the multiple biological effects of β-glucan, it is not surprising that this immunomodulator is also involved in the fight against cancer. Despite the fact that most tumors are recognized by the immune system, the antibody response is usually not strong enough to kill cancer growth.

Since the first direct scientific study forty years ago, the anti-tumor activity of β-glucan has been clearly demonstrated [32]. Since these pioneering studies, numerous animal, as well as human trials, have shown remarkable anti-tumor activity against a wide variety of different tumors including breast, lung, and gastrointestinal cancer. Since the 1980s, two types of β-glucan have been successfully used as a traditional medicine for cancer therapy in Japan and China. In Japan, β-glucan is already licensed as a drug effective in cancer treatment and its treatments are routinely improved. In addition, at least 26 clinical trials are currently under way in the United States as well as several European countries such as Turkey and France.

We investigated the occurrence of antibodies and C3 fragment in both the animal and human models. Our investigation showed that the majority of malignant cells in mammary carcinomas are naturally targeted with C3 for cytotoxicity by NK cells bearing CR3 receptor that has been primed with β-glucan. Both freshly excised human mammary tumors and established breast cancer cell lines were examined, and published reports of both circulating antibodies to tumors and tumor opsonization with immunoglobulins and C3 were confirmed. Further, whereas older investigations had tested tissue sections by immunohistochemistry, our investigation examined single cell suspensions of tumors by flow cytometry, thus allowing full quantification of antibodies and C3 fragment of complement [33]. Our results suggested that while the majority of malignant cells within tumors bore IgM, IgG and C3, the surrounding normal breast epithelium was devoid of these immune reactants. Similarly, it was reported that tumor section from all 48 tested patients with mammary carcinoma were positive for IgG as well as C3 [34].

Numerous recent studies have shown that β-glucan is extremely active in cooperation with antibodies that naturally occur in case of cancer. Experiments that followed showed that daily therapy with soluble or insoluble β-glucan for two weeks resulted in a 70 to 95 % reduction in tumor weight as compared to the control group [14, 35]. We have to keep in mind that antibodies alone cannot make tumor cells disappear. However, following the binding of antibodies on the surface of cancer cells, C3 fragments of complement coat the cancer cells. The β-glucan-primed cells, such as blood neutrophils, macrophages, and natural killer cells, then specifically recognize these complement-antibody complexes and kill the tumor cells. β-Glucan cooperation with antibodies belongs therefore to the most promising combination therapies (for review see [36 - 38]).

Cancer patients observed additional benefits of β-glucan. A recently published study by the research teams from Germany and Turkey found that β-glucan induces proliferation and activation of monocytes in peripheral blood of patients with advanced breast cancer [39]. These findings indicate that β-glucan not only helps the bone marrow to overcome negative effects of cancer, chemotherapy, and irradiation, but also increases the biological activities of these newly-formed immunocytes.

Effects of β-Glucans on Blood Sugar and Cholesterol

In addition to the effects of β-glucan oriented towards the immune system, cereal β-glucans were also shown to reduce the total and LDL cholesterol levels of hypercholesterolemic animals and patients. Fifty years ago, the possible effects of dietary fiber were first suggested by Keys [40] and these effects were later found to be associated with β-glucans [41]. The cholesterol-lowering effects of fibers are routinely associated with β(1-4) glucans. Due to the high consumption of oats or oat bran, most of the attention has been focused on the relationship of oat-derived β-glucan to cholesterol levels in both animals and humans. These β-glucans were shown to reduce serum cholesterol in both hypercholesterolemic animals [42 - 44] and humans [45], however there are some studies that found no such effects [46].

In addition, fiber-containing β-glucans, particularly the ones in human food, gained attention for their role in the metabolic control of diabetes [47 - 49]. The cholesterol-lowering effects of β-glucans are well-established on numerous models, including humans [43, 50]. Without any direct proof, these effects are usually described as the result of fiber intake and subsequent decreased absorption of bile acids. However, most of these studies suffer from the fact that they did not evaluate the effects of isolated β-glucans. They only used crude extracts [51] without any knowledge if these β-glucans are even digested.

Most of experimental studies dealing with cholesterol-lowering effects of β-glucans used oats- or barley-derived β-glucans [43, 44], without finding significant differences between these β-glucans [42]. We compared several commercially important β-glucans and evaluated their ability to lower cholesterol. First, we studied the effect of long-term feeding with β-glucan-added diet. Our data showed strong time-dependent effects of #300 and Krestin β-glucans on lowering cholesterol. Effects of other β-glucans were less pronounced and in the case of ImmunoFiber we found almost no effects (except after 50 days of feeding).

The mice were then given a diet with added cholesterol. The blood cholesterol levels obtained after two weeks of cholesterol feeding were used as positive control. The cholesterol-rich diet was followed by 40 days of feeding with β-glucan-rich diet. Individual groups of mice were sacrificed in 10-day-intervals and cholesterol levels were evaluated. Results showed that during short time intervals all β-glucans lowered the cholesterol levels in hypercholesterolemic animals, but in long term only #300 β-glucan retained this activity.

The current study is not only the first to directly compare the cholesterol-lowering activity of two different yeast-derived β-glucans, but also the first to compare normal animals and mice with experimentally-induced cholesterolemia [52]. In addition, we showed that the type of branching is probably not responsible for these effects.

The effects of β-glucans on blood sugar levels are less known. Some studies showed hypoglycemic activity of natural β-glucans [48, 53], additional studies demonstrated strong hypoglycemic activity of synthetic polysaccharides [54, 55]. However, the mechanisms remain unknown. Some groups suggested similar mechanisms as in lowering cholesterol, *i.e.,* changes in the increase of viscosity of the alimentary bolus and changes in the gastric emptying [56]. Our data found no correlation between these effects and source of the β-glucan or its solubility.

We tested the same β-glucans as mentioned above on effects of β-glucan administration on levels of blood sugar. Feeding with β-glucan did not significantly affect the sugar levels. However, a different situation was found when we used mice with experimentally - induced hyperglycaemie. After two weeks of feeding, #300 β-glucan significantly lowered the sugar levels to almost normal. A longer application of β-glucan resulted in additional significant activity of ImmunoFibre β-glucan [57, 58].

Similar activity was also found with yeast-derived β-glucan *Betamune*. The fact that both β-glucans are insoluble indicates indicates the lack of relevance of the solubility. The fact that both soluble and insoluble β-glucans are processed by macrophages into small fragments that subsequently prime cells for biological activity [14], further stresses the low (if any) role of β-glucan solubility. In addition, the effects of both types of β-glucans suggest that the general mechanism is most probably acting *via* increased intestinal viscosity, causing the reduction of cholesterol absorption, and ending with its subsequent excretion.

Other Biological Effects

In addition to the effects mentioned above, β-glucan has been shown to have other numerous biological effects. An interesting study used β-glucan for the treatment of patients with allergic rhinitis. The results of the study showed that levels of interleukin 4 and 5, which are responsible for the allergic inflammatory response, were decreased with β-glucan treatment, while the levels of interleukin 12 were increased. Moreover, the eosinophils, which are important effector cells of the inflammatory response, were decreased. In summary, β-glucan might have a role as an adjunct to the standard treatment of patients with allergic rhinitis [59].

β-Glucan was recently found to have an additional use-the regulation of stress. We measured the effects of various types of β-glucan on the levels of stress-induced corticosterone. As experimentally induced stress, we used either restraint or cold. Our results [60] showed that β-glucans successfully helped to keep the stress hormone corticosterone at almost normal levels (Fig. **1**). An additional group tested a different type of stress—the oxidative stress of hepatocytes. Evaluating the genotoxic and cytotoxic effects of various substances on freshly isolated hepatocytes, the protective effects of β-glucans were demonstrated [61].

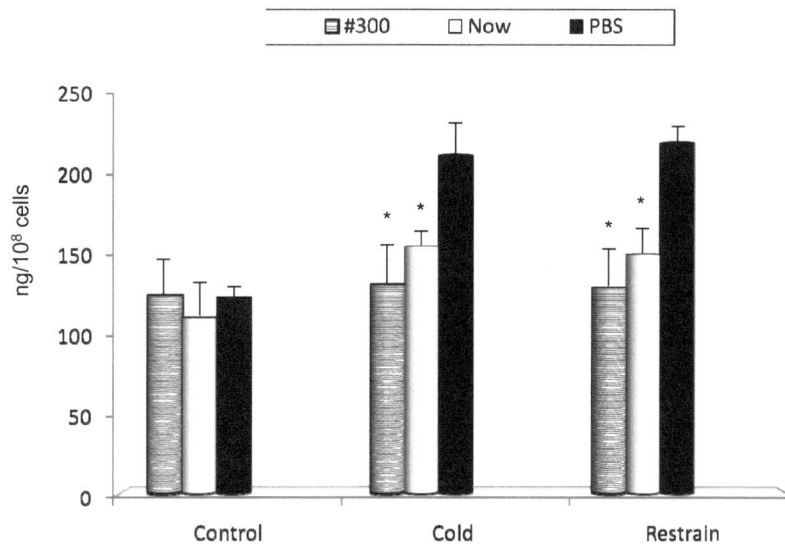

Figure 1: Effect of 14 days of β-glucan feeding on stress-induced levels of corticosterone. *Represents significant differences between stressed-control (PBS) and β-glucan samples at P ≤0.05 level.

Studies showing the strong potential of β-glucans to help overcome immunosuppressive effects of factors, such as irradiation or chemotherapy, led us to the hypothesis evaluated in this paper. The aims of the present study are to compare immunosuppression caused by either organic (thimerosal) or inorganic (mercury acetate) mercury and to show if this suppression can be reversed by β-glucan [62]. We found that two weeks of a daily dose of mercury acetate corresponding to approximately 800 μg of mercury/kg (or 200 μg of mercury/kg in case of thimerosal) induced a systemic suppression of all tested immune reactions, from cellular (phagocytosis, NK cell activity, mitogen-induced proliferation and expression of CD markers) to humoral immunity (antibody formation and secretion of IL-6, IL-12 and IFN-γ).

The mice were then fed with a diet containing a standard dose of β-glucan. Our results showed that simultaneous treatment with mercury and β-glucan resulted in significantly lower immunotoxic effects of mercury which suggests that β-glucans can be successfully used as a natural remedy of low level exposure to mercury. Out of four tested β-glucans, only one showed significant effects on mercury-suppressed LPS-induced proliferation of B lymphocytes (Fig. **2**).

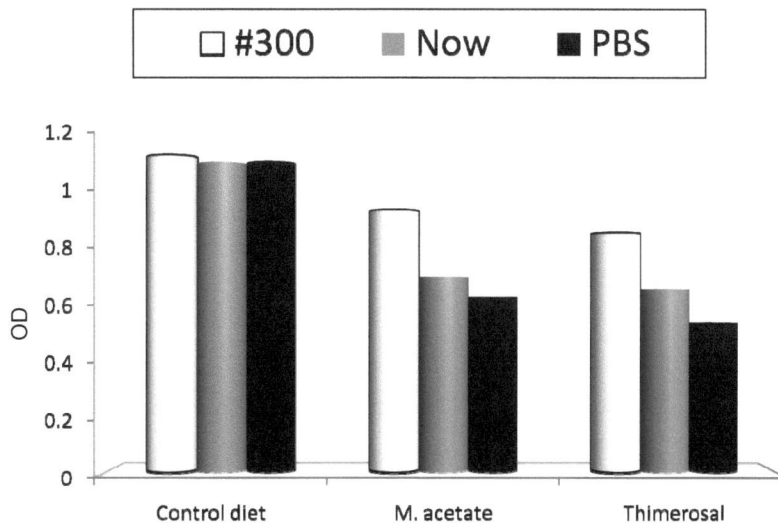

Figure 2: Effect of Hg compounds and different β-glucan samples on LPS-induced proliferation of B lymphocytes.

Regarding the mechanisms by which β-glucan reverses Hg-mediated immunosuppression, the answer remains unclear. In addition to direct stimulation of cells *via* Dectin-1 and CR3 receptors, β-glucans are known to alter some important genes and their transcription factors (for review see [63]). Moreover, since Hg causes inflammation and oxidative stress [64], intracellular mechanisms that involve antioxidant processes might be assumed as well [65].

CONCLUSION

In conclusion, it is well established that among the many known and tested immunomodulators, polysaccharides isolated from various natural sources occupy a prominent position. Due to their very low (if any) toxicity, there was a time when β-glucans were considered merely a matter of fashion. Effects of β-glucans on a variety of diseases, such as infections, immunosuppression, and foremost on neoplastic growth, were investigated. A flood of various food additives and "alternative remedies," usually offered by faintly enlightened non-specialists, is a holdover from these pioneer years. Not surprisingly, a wave of enthusiasm slowly fell off. β-Glucans were, and often incompetently, criticized by many authorities, the main reasons being insufficiently defined preparations and non-specific and/or complex effects.

Fortunately, in the last years, research in reputable laboratories has reached a phase when the basic mechanisms of β-glucan effects are known and the relationship between structure and activity were clearly established. It seems now that β-glucans will finally take a position which was ascribed to them more than fifty years ago.

ACKNOWLEDGMENT

The work of Miroslav Novak was supported by the grant of the Czech Science Foundation No. 525/09/1033

REFERENCES

[1]   Czop JK, Austen KF. Properties of glycans that activate the human alternative complement pathway and interact with the human monocyte β-glucan receptor. J Immunol 1985; 135: 3388-93.

[2]   Patchen ML, MacVittie TJ. Use of glucan to enhance hemopoietic recovery after exposure to cobalt-60 irradiation. Adv Exp Med Biol 1982; 155: 267-72.

[3]   Springer TA. Traffic signals for lymphocyte recirculation and leukocyte emigration: the multistep paradigm. Cell 1994; 76: 301-14.

[4]   Petty HR, Todd RF. Receptor-receptor interactions of complement receptor type 3 in neutrophil membranes. J Leukocyte Biol 1993; 54: 492-4.

[5]   Sutterwala FS, Rosenthal LA, Mosser DM. Cooperation between CR1 (CD35) and CR3 (CD 11b/CD18) in the binding of complement-opsonized particles. J Leukocyte Biol 1996; 59: 883-90.

[6]   Janusz MJ, Austen KF, Czop JK. Isolation of a yeast heptaglucoside that inhibits monocyte phagocytosis of zymosan particles. J Immunol 1989; 142: 959-65.

[7]   Diamond MS, Garcia-Aguilar J, Bickford JK, Curbi AL, Springer TA. A subpopulation of Mac-1 (CD11b/CD18) molecules mediates neutrophil adhesion to ICAM-1 and fibrinogen. J Cell Biol 1993; 120): 1031-43.

[8]   Balsam LB, Liang TW, Parkos CA. Functional mapping of CD11b/CD18 epitopes important in neutrophil - epithelial interactions: a central role of the I domain J Immunol 1998; 160: 5058-65.

[9]   Forsyth CB, Plow EF, Zhang L. Interaction of the fungal pathogen *Candida albicans* with integrin CD11b/CD18: recognition by the I domain is modulated by the lectin-like domain and the CD18 subunit. J Immunol 1998; 161: 6198-205.

[10]   Ross GD, Vetvicka V, Thornton BP. In Phagocyte Functions: A Guide for Research and Clinical Evaluation, Wiley, New York, 1998; pp. 1-17.

[11]   Xia Y, Ross GD. Mapping the β-glucan-binding lectin site of human CR3 (CD11b/CD18) with recombinant fragments of CD11b. FASEB J 1998; 12: A907.

[12]   Schepetkin I A, Quinn MT. Botanical polysaccharides: macrophage immunomodulation and therapeutical potential. Int. Immunopharmacol 2006; 6: 317-33.

[13]   Vetvicka V, Dvorak B, Vetvickova J, *et al.* Orally-administered marine (1-3)-β-D-β-glucan Phycarine stimulates both humoral and cellular immunity. Int J Biol Macromol 2007; 40: 291-8.

[14]   Hong F, Yan J, Baran J T, *et al.* Mechanism by which orally administered β-glucans enhance the tumoricidal activity of antitumor monoclonal antibodies in murine tumor models. J Immunol 2004; 173: 797-806.

[15] Chan GCF, Chan WK, Sze DMY. The effects of β-glucan on human immune and cancer cells. J Hematol Oncol 2009; 2: 25-35.

[16] Al Tuwaiji AS, Mahmoud AA, Al Mofleh IA, Al Khuwaitir SA. Effect of glucan on *Leishmania major* infection in BALB/c mice. J Med Microbiol 1987; 23: 363-5.

[17] Cook JA, Holbrook TW. Immunogenicity of soluble and particulate antigens from *Leishmania donovani*: effect of glucan as an adjuvant. Infect Immun 1984; 40: 1038-43.

[18] Bacon JSD, Farmer VS. The presence of predominantly β-(1-6)-component in preparations of yeast glucan. Biochem J 1968; 110: 34P-5P.

[19] Bousquet M, Escoula L, Pipy B, Bessieres MH, Chavant L, Seguela JP. Two β1-3, β-1-6 polysaccharides (PSAT and Scleroglucan) enhance the resistance of mice to *Toxoplasma gondii*. Ann Parasitol Hum Comp 1988; 63: 398-409.

[20] Dritz SS, Shi J, Klelian TL, *et al.* Influence of dietary β-glucan on growth performance, non-specific immunity, and resistance to *Streptococcus suis* infection in weanling pigs. J Anim Sci 1995; 73: 3341-50.

[21] Kumar P, Ahmad S. Glucan-induced immunity in mice against *Plasmodium bergei*. Ann Trop Med Parasitol 1985; 79: 211-3.

[22] Liang J, Melican D, Cafro L, *et al.* Enhanced clearance of a multiple antibiotic resistant *Staphylococcus aureus* in rats treated with PGG-glucan is associated with increased leukocyte counts and increased neutrophils oxidative burst activity. Int J Immunopharmacol 1998; 20: 595-614.

[23] Rasmussen TL, Seljelid R. Dynamics of blood compontents and peritoneal fluid during treatment of murine *E. coli* sepsis with β-1,3-D-polyglucose derivates. I. Cells. Scand J Immunol 1991; 31: 321-31.

[24] White TR, Thompson RC, Penhale WJ, Chihara G. The effects of lentinan on the resistance of mice to *Mesocestoides corti*. Parasitol Res 1988; 74: 563-8.

[25] Williams DL., Yaeger RG, Pretus HA, Browder IW, McNamee RB, Jones EL. Immunization against *Trypanosoma cruzi*: adjuvant effect of glucan. Int J Immunopharmacol 1989; 11: 403-10.

[26] Yun CH, Estrada A, Van Kessel A, Gajadhar A, Redmond M, Laarveld B. Immunomodulatory effects of oat β-glucan administered intragastrically or parenterally on mice infected with *Eimeria vermiformis*. Microbiol Immunol 1998; 42: 457-65.

[27] Větvička V, Terayama K, Mandeville R, Brousseau P, Kournikakis B, Ostroff G. Pilot study: orally administered yeast β1,3-glucan prophylactically protects against anthrax infection and cancer in mice. J Am Nutraceut Assoc 2002; 5: 1-6.

[28] Babineau TJ, Marcello P, Swalis W, Kenler A, Bistrian B, Forse RA. Randomized phase I/II trial of a macrophage - specific immuno-modulatory (PGG-β-glucan) in high-risk surgical patients. Annals of Surgery 1994; 220: 601-9.

[29] Browder W, Williams D, Pretus HA, Enrichsen F, Mao P, Franchello A. Beneficial effect of enhanced macrophage function in the trauma patients. Ann Surg 1990; 211: 605-13.

[30] Sener G, Toklu H, Ercan F, Erkanh G. Protective effect of β-glucan against oxidative organ injury in a rat model of sepsis. Int Immunopharmacol 2005; 5: 1387-96.

[31] Yang C, Gao J, Dong H, Zhu PF, Wang ZG, Jiang JX. Expression of scavenger receptor, CD14 and protective mechanisms of carboxymethyl-β-1,3 glucan in posttraumatic endotoxemia in mice. J Trauma 2008; 65: 1471-7.

[32] Nakao I, Uchino H, Kaido I, *et al.* Clinical evaluation of schizophyllan (SGP) in advanced gastric cancer – a randomized comparative study by an envelop method. Jpn J Cancer Chemotherap 1983; 10: 1146-59.

[33] Vetvicka V, Thornton BP, Ross GD. Soluble β-glucan polysaccharide binding to the lectin site of neutrophil or natural killer cell complement receptor type 3 (CD11b/CD18) generates a primed state of the receptor capable of mediating cytotoxicity of iC3b-opsonized target cells. J Clin Invest 1996; 98: 50-61.

[34] Niculescu F, Rus HG, Retegan M, Vlaicu R. Persistent complement activation on tumor cells in breast cancer. Am J Pathol 1992; 140: 1039-43.

[35] Hong F, Hansen RD, Yan J, *et al.* β-glucan functions as an adjuvant for monoclonal antibody immuno-therapy by recruiting tumoricidal granulocytes as killer cells. Cancer Res 2003; 63: 9023-31.

[36] Liu J, Gunn L, Hansen R, Yan J. Yeast-derived β-glucan in combination with anti-tumor monoclonal antibody therapy in cancer. Recent Patent on Anti-Cancer Drug Discovery, 2009a; 4: 101-9.

[37] Liu J, Gunn L, Hansen R, Yan J. Yeast-derived β-glucan with anti-tumor monoclonal antibody for cancer immunotherapy. Exp Mol Pathol 2009b; 86: 208-14.

[38] Zhong W, Hansen R, Li B, *et al.* Effect of yeast-derived β-glucan in conjuction with Nevacizumab for the treatment of human lung adenocarcinoma in subcutaneous and orthotopic xenograft models. J Immunother 2009; 32: 703-12.

[39] Demir G, Klein HO, Mandel-Molinas N, Tuzuner N. Beta glucan induces proliferation and activation of monocytes in peripheral blood of patients with advanced breast cancer. Internal Immunopharm 2007; 7: 112-6.

[40] Keys A, Anderson JT, Grande F. Diet-type (fast constant) and blood lipid in man. J Nutr 1960; 70: 257-66.

[41]   Tietyen JL, Nevins DJ, Schneeman BO. Characterization of the hypocholesterolemic potential of oat brand. FASEB J 1990; 4: A527.

[42]   Delaney B, Nicolosi RJ, Wilson TA, Carlson T, Frazer S, Zheng GH, *et al.* Beta-β-glucan fractions from barley and oats are similarly antiatherogenic in hypercholesterolemic Syrian golden hamsters. J Nutr 2003; 133: 468-75.

[43]   Fadel JG, Newman RK, Newman CW, Barnes AE. Hypocholesterolemic effects of beta-β-glucans in different barley diets fed to broiler chicks. Nutr Rep Int 1987; 35: 1049-58.

[44]   Kahlon TS, Chow FI, Knuckles BE, Chiu MM. Cholesterol-lowering effects in hamsters of β-glucan - enriched barley fraction, dehulled whole barley, rice bran, and oat bran and their combinations. Cereal Chem 1993; 70: 435-40.

[45]   Queenan KM, Stewart ML, Smith KN, Thomas W, Fulcher RG, Slavin JL. Concentrated oat β-glucan, a fermentable fiber, lowers serum cholesterol in hypercholesterolemic adults in a randomized controlled trial. Nutrition J 2007; 6: 1-8.

[46]   Keogh GF, Cooper GJ, Mulvey TB, *et al.* Randomized controlled crossover study of the effect of a highly β-glucan-enriched barley on cardiovascular disease risk factors in mildly hypercholesterolemic men. Am J Clin Nutr 2003; 78: 711-18.

[47]   Wursch P, Pi-Sunyer FX. The role of viscous soluble fiber in the metabolic control of diabetes: a review with special emphasis on cereals rich in beta-β-glucan. Diabetes Care 1997; 20: 1774-80.

[48]   De Paula ACCFF, Sousa RV, Figueiredo-Ribeiro, RCL, Buckeridge MS. Hypoglycemic activity of polysaccharide fractions containg β-glucans from extracts of *Rhynchemytrum repens* (Willd.) C.E. Hubb., Poaceae. Braz J Med Biol Res 2005; 38: 885-93.

[49]   Hong L, Xun M, Wutong W. Anti-diabetic effects of β-glucan from fruit body of maitake (*Grifola frondosa*) on KK-Ay mice. J Pharmacy Pharmacol 2007; 59: 575-82.

[50]   Reyna-Villasmil N, Bermudez-Pirela V, Mengual-Moreno E, *et al.* Oat-derived β-glucan significantly improves HDLC and diminishes LDLC and non-HDL cholesterol in overweight individuals with mild hyper-cholesterolemia. Am J Therapeut 2007; 14: 203-12.

[51]   Wang L, Behr SR, Newman RK, Newman CW. Comparative cholesterol-lowering effects of barley β-glucan and barley oil in golden syrian hamsters. Nutr Res 1997; 17: 77-88.

[52]   Vetvicka V, Vetvickova J. Effect of yeast-derived β-glucans on blood cholesterol and macrophage functionality. J Immunotoxicol 2009; 6: 30-5.

[53]   Battilana P, Ornstein K, Minehire J, *et al.*, Mechanisms of action of β-glucan in postprandial glucose metabolism in healthy men. Eur J Clin Nutr 2001; 55: 327-33.

[54]   Hatanaka K, Song SC, Maruyama A, *et al.* A new synthetic hypoglycaemic polysaccharide. Biochem Biophys Res Commun 1992; 188: 16-19.

[55]   Sone Y, Makino C, Misaki A. Inhibitory effects of oligosaccharides derived from plant xyloβ-glucan on interstinal glucose absorption in rat. J Nutr Sci Vitaminol 1992; 38: 391-5.

[56]   Wood PJ. Physicochemical properties and physiological effects of the (1-4) (1-4)-beta-D-glucan from oats. Adv Exp Med Biol 1990; 270: 119-27.

[57]   Vetvicka V, Vetvickova J. A comparison of injected and orally administered β-glucans. J Amer Nutrit Assoc 2008; 11: 41-9.

[58]   Vetvicka V, Vashishta A, Saraswat-Ohri S, *et al.* Immunological effects of yeast- and mushroom-derived β-β-glucans. J. Medicinal Foods 2008; 11: 615-22.

[59]   Kirmaz C, Bayrak P, Yilmaz O, Yuksel H. Effect of β-glucan treatment on the Th1/Th2 balance in patients with allergic rhinitis: a double-blind placebo-controlled study. Eur Cytokine Netw 2005; 16: 128-34.

[60]   Vetvicka V. Vancikova, Z. Antistress action of several glucans. Food Agri Immunol in press.

[61]   Horvatova E, Eckl PM, Bresgen N, Slamenova D. Evaluation of genotoxic and cytotoxic effects of H_2O_2 and DMNQ on freshly isolated rat hepatocytes; protective effects of carboxymethyl chitin - β-glucan. Neuroendocrinol Lett 2008; 29: 644-8.

[62]   Vetvicka V, Vetvickova J. Effects of glucan on immunosuppressive actions of mercury. J Medicinal Food 2009; 12: 1098-1104.

[63]   Novak M, Vetvicka V. β-glucans, history and the present: Immunomodulatory aspects and mechanisms of action. J Immunotoxicol 2008; 5: 47-57.

[64]   Sarafian TA. Methylmercury-induced generation of free radicals: biological implications. Met Ions Biol Syst 1996; 36: 415-44.

[65]   Quig D. Cysteine metabolism and metal toxicity. Altern Med Rev 1998; 3: 262-70.

Biology and Chemistry of Beta Glucan, Vol. 01, 2011, 19-38

CHAPTER 3

β-Glucan Receptors

Alexandra E. Clark[1], Ann M. Kerrigan[1] and Gordon D. Brown[1]*

[1]Section of Immunology and Infection, Institute of Medical Sciences, University of Aberdeen, Aberdeen, UK

Abstract: β-Glucans are naturally occurring carbohydrates found in plants, fungi and some bacterial species. They possess immunomodulating activities, but their mechanisms of action are not well understood. β-Glucans are recognised by the vertebrate immune system *via* pattern recognition receptors such as Dectin-1. Recent studies of Dectin-1 and other β-glucan receptors have enabled us to begin to unravel the mechanisms underlying the immune stimulating properties of these carbohydrates. There is also a strong evidence that β-glucan recognition is an important component of anti-fungal immunity. Here we will review the mammalian β-glucan receptors, with a particular focus on Dectin-1.

INTRODUCTION

The vertebrate immune system has evolved to detect and respond to the vast array of microbes which it encounters on an ongoing basis. The innate immune system is the first line of host defence and its underlying mechanisms can occur within minutes of encountering a potentially threatening microbe. It results in a variety of responses including some, such as the production of cytokines and chemokines, and the presentation of microbial antigens to lymphocytes, which trigger the adaptive arm of the immune system [1]. The adaptive arm of the immune system relies on lymphocytes which recognise antigens from specific pathogens *via* their antigen receptors with the resultant triggering of clonal expansion, cellular differentiation and the production of specific antibodies. The adaptive immune system is also responsible for developing immunological memory which allows it to respond more quickly and efficiently to recurrent infections.

The innate immune system is composed mainly of phagocytic cells and its ability to respond immediately to infection is mediated by the expression of evolutionarily conserved germ-line encoded receptors on these cells, the pattern recognition receptors (PRRs) [2]. These receptors recognise highly conserved structures found in most microorganisms, so-called pathogen-associated molecular patterns (PAMPs) [3]. Contrary to their name, PAMPs are not necessarily unique to pathogens but may also be present on commensal microorganisms [4]. Recent evidence has also demonstrated that PRRs recognise endogenous ligands which are released from or are associated with damaged or dying cells, known as damage-associated molecular patterns (DAMPs) [5, 6].

The best characterised PRRs are the Toll-like receptors (TLRs) which are transmembrane proteins expressed on the cell surface or intracellularly within vacuoles. The TLRs sense various PAMPs such as lipopolysaccharide (LPS), DNA, RNA, lipopeptides and flagellin and they initiate immune responses which have been reviewed extensively elsewhere [1, 7, 8]. Intracellular cytosolic PRRs include the nucleotide-oligomerisation domain (Nod)-like receptors and the retinoic acid-inducible gene-1 (RIG-1)-like receptors which are involved primarily in the detection of bacterial and viral PAMPs respectively [9-11]. C-type lectin receptors (CLRs) comprise transmembrane and soluble proteins which can detect a broad range of mostly carbohydrate components on pathogens [12-14]. Akin to the TLRs, the NLRs, RLRs and some CLRs have been shown to induce innate immune responses and intracellular signalling resulting in triggering of adaptive immunity [15]. Activation of other PRRs, such as the scavenger receptors and the mannose receptor (a CLR), by microbial ligands can initiate uptake and killing but do not trigger the development of adaptive immunity by themselves.

Fungi express a variety of PAMPs whose detection by the innate immune system has been attributed to several PRRs [16, 17]. In this chapter we will focus on the PRRs which recognise β-glucans, carbohydrate PAMPs that are predominantly found in fungal cell walls, but are also present in plants and some bacteria. These naturally occurring

*Address correspondence to: Professor Gordon Brown, Section of Infection and Immunity, Institute of Medical Sciences, School of Medicine and Dentistry, University of Aberdeen, Aberdeen, AB25 2ZD, UK. Email: gordon.brown@abdn.ac.uk

Vaclav Vetvicka and Miroslav Novak (Eds)

polysaccharides are a heterogeneous group of glucose polymers. In fungi they consist primarily of long linear β(1-3)-linked backbones with β(1-6)-linked side chains of varying length and distribution. β(1-3)-Glucan chains have a coiled spring-like structure and interchain hydrogen bonds enable the formation of triple helices, which form a complex meshwork providing structural support and elasticity to the fungal cell wall [18-21].

The immune stimulating properties of β-glucans were first discovered in the 1960's [22]. Since then, and particularly more recently with the massive expansion of the health food industry, commercial preparations of β-glucans sold as food supplements have been extolled for their immune boosting powers. Although there is now a substantial amount of literature concerning the effects of these carbohydrates on the immune system, much of the older work is confusing and frequently contradictory. The undertaking of studies using different cell types and model systems has somewhat complicated interpretations. However the inconsistencies mostly stem from the use of heterogeneous β-glucans which vary in their origin, purity and molecular structure, factors which are reported to influence their bioactivity [23-26]. We now know that physical properties of β-glucans such as tertiary structure, polymer length and the degree of branching can influence their binding affinities for receptors [27, 28]. Although we do not fully understand how binding differences translate into differences in immune modulating activity, we can make some generalizations concerning their cellular effects and molecular weight. Large particulate β-glucans, such as zymosan and curdlan, can directly activate leukocytes stimulating phagocytosis and the production of inflammatory mediators and reactive oxygen intermediates (for examples see [29-32]). The effects of intermediate sized β-glucans, such as glucan phosphate, are not as clear. Although these carbohydrates seem to possess biological activity *in vivo* [33], they do not generally appear to directly activate leukocytes as evidenced by *in vitro* studies. Nevertheless there is evidence that in some instances they can induce cytokines and transcription factors [34-36]. Small β-glucans such as laminarin are not biologically active, although they can still bind to β-glucan receptors and can function as antagonists [27, 37].

The identification and study of PRRs that recognise β-glucans, advances in the understanding of the complexity of β-glucan structure, and standardised protocols for their purification, have enabled significant advances in our understanding of the mechanisms involved in the recognition and response to these carbohydrates. Complement receptor 3 (CR3) was identified as a β-glucan receptor almost 25 years ago [38]. However, a number of other β-glucan receptors have also been recently identified, namely Dectin-1, langerin, lactosylceramide and the scavenger receptors CD5, CL-P1, SCARF1 and CD36 (Fig. 1)

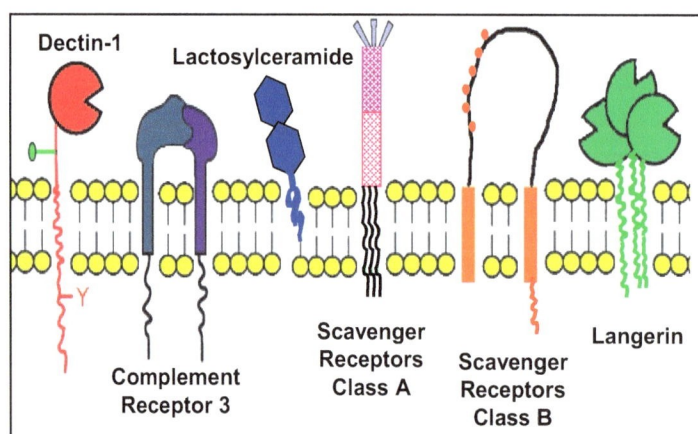

Figure 1: Schematic diagram of β-glucan receptors.

IMMUNOMODULATING PROPERTIES OF β-GLUCANS

The immunomodulating properties of β-glucans have prompted investigations of their use in a number of disease scenarios. A detailed review of such studies is beyond the scope of this chapter, however we will summarise briefly some of the beneficial applications that have been reported. The administration of β-glucans has been shown to have anti-infective effects in animal models of bacterial, viral, fungal and parasitic disease (for review see [39, 40]). These effects of β-glucans are believed to arise from their direct stimulation of leukocytes as mentioned above, but the precise mechanisms by which they mediate their anti-infective activities are not fully understood. β-Glucans are nonetheless promoted as immune boosters and included as feed supplements in farmed pigs, poultry and fish [41-44]. Furthermore, clinical trials have shown that β-glucan administration can reduce serious postoperative infections in patients who have undergone high-risk gastrointestinal procedures [45].

In addition to their anti-infective properties, β-glucans have been of intense interest for their anti-tumor effects. The first reports of their anti-tumor activity emerged over 50 years ago [46]. Since then several human clinical trials have shown possible therapeutic benefits, although results have been variable and unpredictable [47]. Nevertheless, a number of β-glucans have been approved for clinical use in human cancer treatment in Japan [40]. With regard to the use of β-glucans as cancer therapeutics, the most promising avenue of research lies in their co-administration with anti-tumor monoclonal antibodies [48-50]. This approach has been shown to suppress tumor development and increase survival in a variety of cancer models including human carcinoma xenograft models [51-57]. The mechanisms thought to underlie these effects involve the β-glucan promotion of complement dependent cellular cytotoxicity and CR3 [54] and are discussed in detail elsewhere in this volume.

β-Glucans have also been reported to have a number of beneficial applications with regard to cardiovascular health. The cholesterol lowering effects of oats were first described in the 1960's and since then, a significant number of investigations have attributed this effect to the cereals β-glucan component [58, 59]. There is limited evidence suggesting that the consumption of β-glucans lowers blood pressure in obese people [60]. A number of studies have also demonstrated that β-glucans are effective in lowering glucose levels in both healthy and diabetic individuals [61] Investigations have suggested that these effects are essentially due to a delayed and somewhat reduced carbohydrate absorption from the intestine [62].

In comparison with the volume of studies reporting beneficial effects of β-glucans, demonstrations of negative side effects are limited. Some early reports showed that the systemic administration of a microparticulate form was associated with hepatosplenomegaly, granuloma formation, microembolisation and enhanced endotoxin sensitivity [63-65]. However these side effects have been overcome with the development of methodologies to convert particulate β-glucans to well-tolerated water forms [66]. Serious side effects were reported when β-glucans were administered to mice in combination with selected non-steroidal anti-inflammatory drugs. This treatment induced gastrointestinal damage resulting in peritonitis and subsequent death [67-70]. An independent investigation which showed no adverse effects following simultaneous treatment with β-glucan and non-steroidal anti-inflammatory drugs has however recently called these results into question [71]. A causal role in the development of respiratory symptoms associated with fungal exposure has been attributed to β-glucans [72, 73]. Although there is a significant body of literature exploring the effects of β-glucans on respiratory health, current epidemiological data is again difficult to interpret. Although some studies suggest that β-glucan exposure has a detrimental effect on respiratory health, other studies have reported no association between the two [73]. Finally, β-glucans have also been shown to trigger rheumatoid arthritis in genetically susceptible mice, suggesting that fungal infection may evoke autoimmune arthritis or other autoimmune conditions in genetically susceptible individuals [74, 75].

β-GLUCAN RECEPTORS

There is no doubt that β-glucans are important in terms of their immunomodulating properties; however as can be deduced from the brief review above, the literature is often confusing and their mechanisms of action are far from fully understood. To successfully exploit the biological effects of these carbohydrates and to improve host defence against fungal pathogens, it is important to continue investigating the receptors involved in β-glucan recognition. A number of mammalian cells have been shown to express β-glucan receptors since their initial identification on monocytes [76]. These comprise both immune and nonimmune cells, including neutrophils, eosinophils, natural killer cells, dendritic cells, endothelial cells, alveolar epithelial cells, fibroblasts, and a variety of macrophages including microglia [66]. As mentioned earlier, there is now a number of specific β-glucan receptors identified. Our group has focused on the investigation of Dectin-1 since we identified it as a β-glucan receptor in 2001. Since then Dectin-1 has emerged as the primary receptor for these carbohydrates, at least on leukocytes [77-80]. We will therefore focus on Dectin-1 for much of this chapter. However, we will also briefly review the other receptors which are likely to play significant roles in non immune cells and whose involvement in regulating immune responses should not be dismissed.

DECTIN-1

Structure

Dectin-1 is a Group V member of the C-type lectin superfamily. It is a type II transmembrane protein consisting of a single extracellular C-type lectin-like domain (CTLD) connected by a stalk to a transmembrane region followed by a short cytoplasmic tail. The CTLD was first recognized as a calcium-dependent carbohydrate-binding domain, but it is also present in non-calcium-dependent protein recognition interactions [81, 82]. The residues normally required for calcium coordination in classical CTLDs are not conserved in Dectin-1. In further contrast to many other C-type lectins, Dectin-1 lacks cysteine residues in its stalk region, suggesting that it probably does not dimerise, and the receptor appears to be expressed and function as a monomer [83].

Dectin-1 is encoded within a cluster of related genes (the Dectin-1 cluster) in the natural killer gene complex on human chromosome 12 and mouse chromosome 6 [12, 84, 85]. Alternative splicing generates at least two isoforms of murine Dectin-1, encoding the full length receptor and a stalkless version (Fig. 2) that exhibit differences in their ability to bind β-glucans and induce cellular responses [86]. Both murine forms are *N*-glycosylated at two sites in the extracellular region [87]. In humans, alternative splicing generates eight isoforms, of which only the two major forms are functional for β-glucan binding [88]. Of the two β-glucan binding human isoforms, the most predominant lacks the stalk region and has no glycosylation sites. The other form contains one *N*-glycosylation site in its stalk region [88, 89] (Fig. 2). Little is known about the functions of the minor isoforms, although one is retained in the cytoplasm and has been shown to associate with a Ran-binding protein, RanBPM [90]. The functional relevance of this interaction remains undetermined at present.

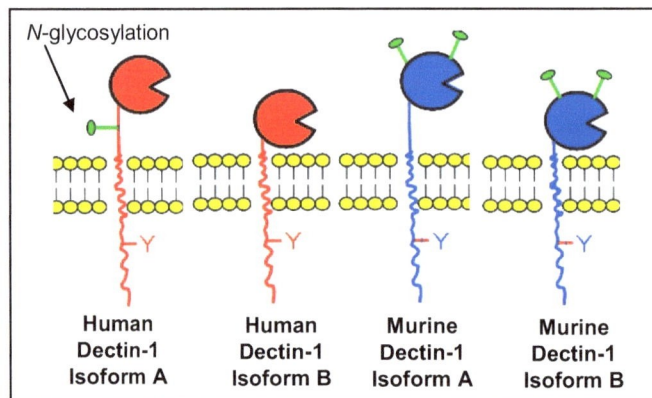

Figure 2: Schematic diagram of Dectin-1 isoforms. Each receptor contains an extracellular C-type lectin domain, a transmembrane region and a cytoplasmic tail. The A isoforms also contain a stalk region. The murine receptors have two *N*-glycosylation sites whilst human isoform A has one *N*-glycosylation site. A tyrosine (Y) residue is found within the cytoplasmic tail which becomes tyrosine phosphorylated to initiate intracellular signalling.

Expression

Both human and murine Dectin-1 are predominantly expressed on myeloid cells including dendritic cells, monocytes, macrophages and neutrophils as well as on subpopulations of T cells [91]. Dectin-1 is also expressed on human B cells, eosinophils and mast cells and recent reports have also demonstrated expression on murine microglia [88, 92, 93]. Expression levels of Dectin-1 on these cells vary within tissues with the highest levels being found at sites of pathogen entry, such as the lung, gut and spleen [91, 94]. It has also been shown that glycosylation contributes to surface expression of murine Dectin-1 [95]. The two major human isoforms are expressed differently in various cell types [88], and there are also differences in Dectin-1 expression between different mouse strains [86]. Dectin-1 expression can also be regulated by various cytokines and microbial factors [78, 96, 97]. For example interleukin (IL)-4, IL-13 and GM-CSF (Granulocyte-macrophage colony-stimulating factor) cause Dectin-1 expression to be highly upregulated. In contrast IL-10, LPS and dexamethasone cause down-regulation of Dectin-1 expression [96]. Furthermore, the systemic administration of both highly purified β-glucans and *Candida albicans* resulted in an increase in Dectin-1 expression on leukocytes [97, 98]. In contrast, Dectin-1 was decreased with polymicrobial sepsis [98].

Ligands

A variety of endogenous and exogenous ligands have been reported but Dectin-1 is best known for its ability to recognise β-glucans. Dectin-1 was initially identified as a dendritic cell specific receptor that modulated T cell function through recognition of an unknown ligand on T cells [99, 100]. It was re-identified as a receptor for β-glucans following a screen of a murine macrophage cDNA expression library with zymosan, a β-glucan rich extract of *Saccharomyces cerevisiae* [37]. The specificity of Dectin-1 for β-glucans was subsequently investigated using oligosaccharide microarray technology which demonstrated that the receptor specifically binds β(1-3)-linked glucose oligomers in a metal ion independent manner [101]. Recent studies have however shown that Dectin-1 does not recognise all β(1-3)-glucans equally and demonstrated that at least a branched heptasaccharide is required for interaction with Dectin-1 [28]. Furthermore this study found that the presence, degree, and perhaps length of side-chains can also influence binding [28]. The Ca^{2+}-independent mechanism for recognition of carbohydrates by Dectin-1 is still unclear, however at least two residues flanking a shallow groove on the protein surface, Trp^{221} and His^{223}, have been implicated for β-glucan binding [87]. A crystal structure of Dectin-1 has been reported in which a short soaked natural β-glucan is trapped in the crystal lattice [102]. In this study no binding was evident in the putative binding groove, but this was possibly due to protonation of His^{223} in the crystallization conditions and/or because the ligand was too short. Modelling studies have in fact demonstrated that the Trp^{221}/His^{223} surface groove could easily accommodate a β-glucan chain [28]. Moreover it has been speculated that the helical nature of glucan polymers may facilitate this interaction and that the binding grove may have a unique structure that preferentially interacts with branched glucans [28]. Dectin-1 has been shown to interact with a number of fungal species including *Candida, Pneumocystis, Saccharomyces, Aspergillus, Coccidioides* and *Penicillum* by way of its β-glucan specificity [29, 103-110]. Recent studies have also shown that Dectin-1 interacts with various mycobacterium species; however the specific ligand(s) recognised by Dectin-1 on these microbes remain unidentified and is unlikely to be β-glucans which have never been described on mycobacteria [111-114]. Finally, as mentioned previously Dectin-1 can also recognise endogenous ligands and has been shown to bind to T cells through recognition of an unknown ligand [99]. In addition studies using cell lines have shown that Dectin-1 recognises an endogenous ligand on apoptotic cells [115].

Dectin-1 Induced Cellular Responses

Dectin-1 has been shown to induce a variety of cellular responses following β-glucan recognition including ligand uptake through phagocytosis and endocytosis, the respiratory burst, the production of arachidonic acid metabolites, dendritic cell maturation and the induction of numerous cytokines and chemokines, including TNFα, CXCL2, IL-23, IL-6, IL-10 and IL-2 [29, 77, 80, 105, 116-121]. Dectin-1 signalling following β-glucan recognition can also trigger adaptive immunity and was the first non-TLR PRR shown to do so. Dectin-1 activated dendritic cells can 'instruct' the differentiation of T helper 17 (Th17) and T helper 1 (Th1) $CD4^+$ T cells in response to *C. albicans* and *Mycobacterium tuberculosis* [116, 122]. Furthermore, Dectin-1 mediated activation of dendritic cells can convert selected populations of regulatory T cells (T regs) into IL-17 producing T cells that cannot be classified as either T regs or Th17 cells [123]. Dectin-1 has also been implicated in the expansion and function of T regs [31, 124]. As well as triggering $CD4^+$ T cell differentiation, Dectin-1 stimulation can drive $CD8^+$ T cell responses [125]. Finally, studies have shown that antibody-mediated targeting of Dectin-1 resulted in the induction of cytotoxic T lymphocyte (CTL) responses as well as $CD4^+$ T cell and antibody responses [126]. Dectin-1 signalling alone is sufficient for some of these activities; however others, for example the production of inflammatory cytokines may require, or are enhanced by, cooperative signalling with MyD88-coupled TLRs [29, 80, 127]. Furthermore the ability of β-glucans to directly trigger these responses is cell-type dependent [128, 129].

Dectin-1 Signalling Pathways

Dectin-1 mediates the cellular responses listed above primarily by signalling through an immunoreceptor tyrosine-based activation (ITAM)-like motif (also called a hemITAM) in its cytoplasmic tail. The traditional ITAM motif as found in T cell receptors, B cells receptors and Fc receptors, is characterised by a consensus sequence that includes two tyrosines, usually 10-12 amino acids apart: $YxxI/Lx_{(6-12)}YxxI/L$ [130-132]. Receptor engagement leads to tyrosine phosphorylation by Src family kinases, and the phosphorylated sequences provide docking sites for Syk kinases by interacting with the two Src homology 2 (SH2) domains in Syk [131, 132]. Following ligand binding Dectin-1 also becomes tyrosine phosphorylated by Src kinases allowing recruitment of Syk. However in contrast to

ITAM receptors where dually phosphorylated tyrosines are necessary for Syk recruitment, phosphorylation of only the membrane proximal tyrosine is sufficient for Syk association with Dectin-1, even though both SH2 domains are still required [77, 133]. It has subsequently been shown that CLEC-2 and CLEC9A, ITAM-like containing members of the Dectin-1 cluster, signal in a similar fashion [133, 134]. Although the exact nature of the interaction between Syk and Dectin-1 is not yet fully understood; a model proposing that Syk bridges two monophosphorylated molecules has become widely accepted [77, 83, 133, 135] (Fig. **3**).

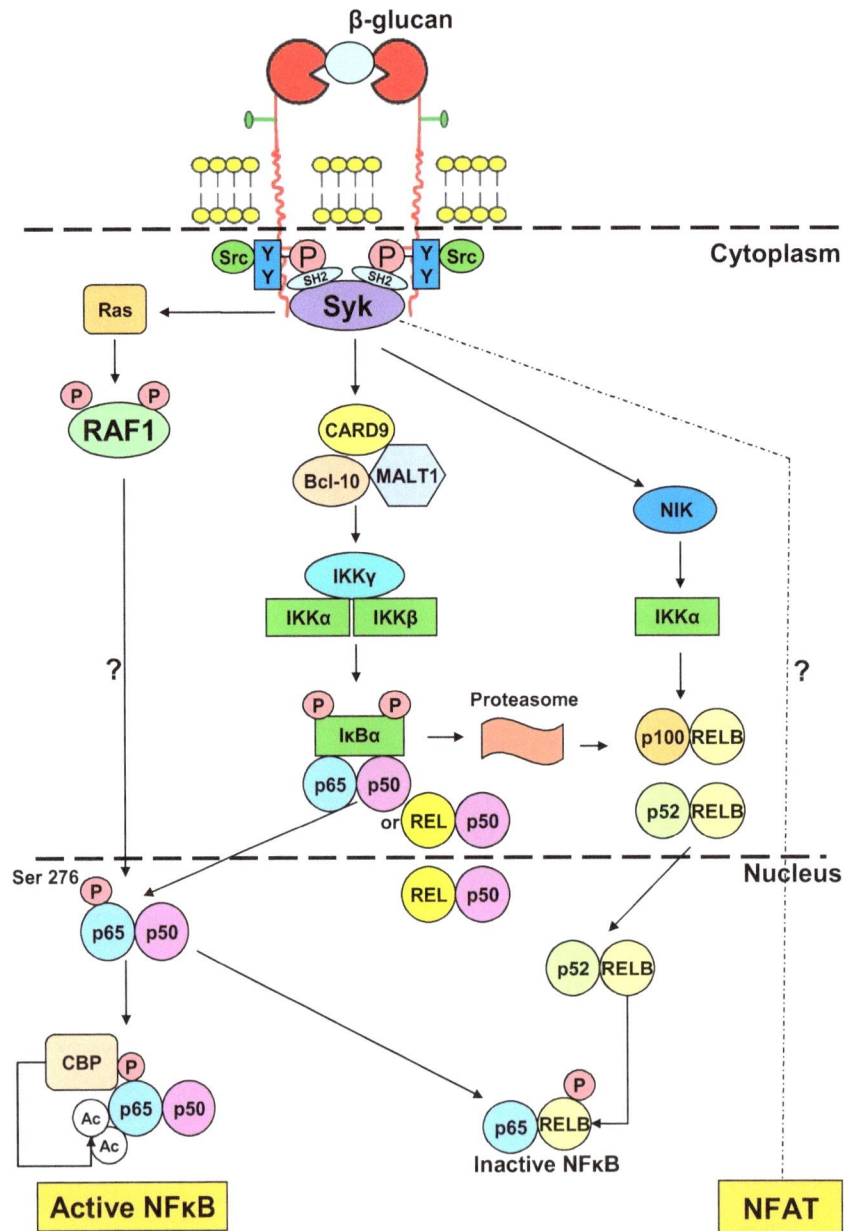

Figure 3: β-glucan ligation of Dectin 1 induces direct activation of NFκB and NFAT (adapted by permission from Macmillan Publishers Ltd: Nature Reviews Immunology, 9(7):465-79, copyright 2009) [147].

Studies in dendritic cells have demonstrated that recruitment of Syk to Dectin-1 results in activation of phospholipase Cγ2 (PLCγ2), leading to the engagement of the caspase recruitment domain (CARD) containing protein, CARD9, which together with Bcl10 (B-cell lymphoma/leukemia 10) and Malt1 (Mucosa-associated lymphoid tissue lymphoma translocation protein 1) signal for activation of the canonical NF-κB pathway [136-138] (Fig. **3**). Precisely how PLCγ2 activation results in CARD9 recruitment is still not known, however protein kinase C

(PKC) may be involved [139]. Dectin-1 is also the first PRR shown to induce Syk dependent activation of NF-κB *via* the non-canonical pathway. [140]. Furthermore Dectin-1 signalling can activate NFAT, although the involvement of Syk in this response has not been demonstrated [138, 141]. There is also evidence that certain cellular responses controlled by Dectin-1/Syk signalling are not dependent on CARD9. For example, in dendritic cells there is Syk-dependent activation of ERK, a mitogen activated protein (MAP)-kinase, through a CARD9-independent pathway [142]. Syk-independent signalling can also be induced by Dectin-1. For example, phagocytosis in macrophages requires the ITAM-like motif and a triacidic DED sequence, but does not require Syk and is possibly propagated by a novel kinase [77, 83, 120]. It has also been shown that Dectin-1 induces a signalling pathway through the kinase Raf-1 that was independent of the Syk pathway but integrated with it at the level of NF-κB activation for regulation of cytokine production [140] (Fig. **3**).

Dectin-1 Collaboration with Other Receptors

An emerging theme in the field of immunology is that microorganisms must stimulate multiple PRRs to induce an effective immune response. Although this makes it much more difficult to form an all-encompassing overview and to design experiments which take all the variables into account, several studies have supported a growing appreciation of this idea [127, 143]. Investigations of Dectin-1 can be included in these studies and it has been shown to collaborate with TLRs for both optimal cytokine production and enhancement of the respiratory burst. In macrophages, collaborative signalling by TLRs and Dectin-1 is required for TNFα production [29, 80]. On the other hand, in DCs, Dectin-1 signalling alone can trigger TNFα and IL-10 production, although collaborative signalling with TLRs enhances these responses [116, 127]. Dectin-1 can in fact interact with several MyD88-coupled TLRs (TLR2, TLR4, TLR5, TLR7, TLR9) to induce production of multiple cytokines cytokines including TNFα, IL-10, IL-6 and IL-23 [127, 144, 145]. Interestingly, these interactions also result in the downregulation of IL-12 [146], and the reciprocal regulation of IL-23 and IL-12 is likely to contribute to the development of Dectin-1 mediated Th17 responses [146].

Dectin-1 is also capable of interacting with other PRRs, namely DC-SIGN and its murine homologue SIGNR1. Costimulation of Dectin-1 and DC-SIGN triggers the arachidonic acid cascade in human dendritic cells [118], and Dectin-1 has been shown to cooperate with SIGNR1 for fungal binding by murine macrophages [148]. In addition to PRRs, Dectin-1 can interact with tetraspanins, a family of proteins that have been shown to modulate signal transduction by interacting with many other transmembrane proteins to form 'tetraspanin microdomains' in the plasma membrane [149]. Dectin-1 associates with the ubiquitous tetraspanin CD63 and it has been shown that phagocytosis of yeast particles by dendritic cells was accompanied by a decrease in CD63 expression, which was inhibitable by a soluble β-glucan, laminarin [150]. The functional significance of a Dectin-1-CD63 interaction remains undefined; however it may represent part of a signalling complex that could influence phagocytosis. Dectin-1 also interacts with the immune cell specific tetraspanin CD37, which co-localises and stabilizes Dectin-1 on the surface of macrophages and mediates control of Dectin-1 induced IL-6 [151, 152]. Whether the tetraspanin-Dectin-1 interactions are involved with Dectin-1 collaboration with TLRs is not known, although there has been speculation that tetraspanins may directly link the Dectin-1 and TLR signalling pathways [151].

Role of Dectin-1 in Anti-Microbial Immunity

β-Glucans can constitute up to 50% of fungal cell walls and as mentioned previously Dectin-1 can recognise a number of different fungal species by way of its β-glucan specificity. Numerous *in vitro* studies have shown that Dectin-1 can induce various protective cellular responses to fungi, including phagocytosis and killing and the production of protective inflammatory cytokines and chemokines such as TNFα, CXCL2, IL-1β, IL-1α, CCL3, GM-CSF, G-CSF and IL-6. Somewhat surprisingly these studies have also shown that Dectin-1 induces the production of IL-2 and IL-10 [77, 107, 116]. The role of these cytokines in fungal infection is less clear. IL-10 is an anti-inflammatory cytokine that has historically been considered as detrimental for the host during fungal infection. However, this thinking has recently been questioned and it is now proposed that the inhibitory action of IL-10 on macrophage and neutrophil activation may contribute to limiting host damage during severe inflammation [153]. IL-10 has furthermore been shown to be involved in the development of T regs which are beneficial in mucosal and cutaneous infections, and are essential components for inducing long term immunity to *C. albicans* [154]. On the other hand, it has been suggested that T regs become detrimental to the host response when a pathogen penetrates the mucosa and disseminates through the bloodstream, as in the case with disseminated candidiasis [154].

Initial studies focused on innate immune responses mediated by Dectin-1, but more recently it has been shown that it can also trigger adaptive immune responses, and indeed it was the first non-TLR PRR shown to do so. As mentioned previously stimulation of the Dectin-1 pathway can drive Th1, Th17, T reg, CTL and antibody responses. These responses have not been specifically studied in the context of Dectin-1 and fungal infection. However, it is generally accepted that a Th1 response is required for efficient anti-fungal immunity [155]. On the other hand, the induction of Th17 cells during systemic fungal infections has been shown to be both beneficial and detrimental to the host [155-158]. Conversely these cells are known to be important during mucosal infection [159]. The specific role of CTLs in direct antifungal responses is similarly unclear, however it has been shown that CD8^{+} T cells are activated during fungal infections and can play a protective role in some cases [160-163].

A number of *in vivo* studies, although not entirely consistent, suggest an important role for this receptor in anti-fungal immunity. Dectin-1 deficient mice exhibited enhanced susceptibility to systemic *C. albicans* infection which correlated with defects in the inflammatory response and a modest defect in fungal killing [107]. These mice also showed increased susceptibility to mucosal candidiasis, with enhanced dissemination and decreased survival times [164]. Mice deficient in CARD9, a signalling component downstream of Dectin-1, were similarly susceptible to systemic *C. albicans* infection [136]. Blocking Dectin-1 *in vivo* after *Aspergillus fumigatus* challenge significantly reduced inflammatory responses and increased fungal burdens in the lung [106]. A subsequent study which challenged Dectin-1 deficient mice with *A. fumigatus* demonstrated that these mice were more susceptible to infection and showed impaired cytokine production, reactive oxygen species production and killing [165]. In contrast to these studies, Dectin-1 deficient mice of a different genetic background did not show defects in cytokine production in response to *C. albicans* or *Pneumocystis carinii* [104]. These mice were not more susceptible to *Candida* infection, but showed increased fungal burdens during *P. carinii* infection which correlated with defective production of reactive oxygen species [104]. A later investigation also used this strain to show that Dectin-1 was not required for host resistance to *Cryptococcus neoformans* [166]. We suspect that the inconsistencies across studies are due to the use of different fungal strains and mice with different backgrounds. Nevertheless, definitive evidence of the role of Dectin-1 in fungal infections is provided by the recent identification of a polymorphism in human Dectin-1. This polymorphism results in defective surface expression of the receptor and is associated with an increased occurrence of mucosal fungal infection that correlates with decreased production of cytokines, including IL-17 [167]. Further evidence is provided by another study which showed that a mutation causing a loss of function of CARD9 in humans, was also associated with a susceptibility to mucosal fungal infections [168].

The traditional view of the architecture of the fungal cell wall is that β-glucans are buried within a matrix of phospholipomannan and mannoproteins below a layer of glucose and chitin [18]. However early studies of Dectin-1 clearly showed recognition of β-glucan containing particles and intact fungi and suggested that these carbohydrates were actually displayed on the cell surface [37]. Subsequent studies have found that Dectin-1 recognises *C. albicans* yeast, but not filamentous forms [109]. These studies also demonstrated that although yeast β-glucans are largely shielded from Dectin-1, the normal mechanisms of yeast budding and cell separation create permanent scars which expose sufficient β-glucan to trigger its activation [109]. Similarly in *A. fumigatus*, swollen conidia and early germlings displaying surface β-glucans are recognised by Dectin-1, hyphal forms are recognised to a lesser degree and resting conidia are not recognised [103, 106, 169, 170]. It has been suggested that fungal pathogens actively mask their β-glucans to avoid immune recognition by Dectin-1[109, 171] and this has prompted investigations into drugs such as caspofungin, which enhance the exposure of β-glucans and improve antifungal responses [171-174].

The Role of Dectin-1 in β-Glucan Induced Immunomodulation

As we have highlighted in this review, β-glucan ligation of Dectin-1 triggers many cellular responses and we think that it is likely that Dectin-1 therefore plays a critical role in the immunomodulatory activities of these carbohydrates. *In vivo* studies demonstrating that defects in Dectin-1 and its downstream signalling components influence the inflammatory and adaptive responses induced by particulate β-glucans provide evidence for this viewpoint [116]. Further support is provided by recent studies which showed that specific blocking of Dectin-1 inhibits β-glucan induced anti-tumor activity [175]. What is not understood yet is why soluble β-glucans such as glucan phosphate are biologically active *in vivo* [36, 176], yet do not induce cellular responses *in vitro* and are, in fact, used as Dectin-1 inhibitors. Further investigations are required to fully appreciate the involvement of Dectin-1 in β-glucan immunomodulation and may perhaps reveal novel therapeutic strategies.

COMPLEMENT RECEPTOR 3 (CR3)

Complement receptor 3 (CR3, Mac-1, $\alpha_m\beta_2$) is part of the β_2-integrin family of adhesion molecules and is a heterodimer composed of CD18 which is common to all β_2-intergrins, and CD11b which is unique to CR3. Its expression has been demonstrated on a wide range of leukocytes including monocytes, macrophages, dendritic cells, neutrophils, eosinophils, natural killer cells and T and B cells. CR3 is remarkable in its ability to recognise diverse ligands and has been shown to bind to several matrix components as well as other endogenous proteins. With regard to immune functions, CR3 recognises the complement component iC3b, on the surface of opsonised pathogens. It also functions as a non-opsonic PRR and recognises a number of microbial ligands, including β-glucan. After an initial demonstration that CR3 bound to unopsonised zymosan, it was subsequently determined that this binding was mediated by β-glucan recognition [38]. The CR3 binding site for iC3b and many other ligands has been localised to a portion of the CD11b subunit known as the I-domain [177, 178]. Its recognition of β-glucans is *via* a distinct lectin site which is located C-terminal to the I-domain of CD11b [179]. CR3 in fact has broad sugar specificity and mediates binding to microbes that do not contain β-glucans, possibly *via* other carbohydrates [179-183].

CR3 has been implicated in mediating some biological effects in response to β-glucans, including phagocytosis, cytotoxicity, the reduction of transendothelial migration, neutrophil chemotaxis, and enhancement of complement-mediated haematopoietic recovery [184-188]. These responses have been primarily studied in neutrophils and it appears that β-glucan binding primes the receptor for enhanced activity. For example, CR3 has been reported to stimulate phagocytosis, degranulation, and the respiratory burst in response to opsonised yeast. However, recognition of β-glucans in addition to complement components is required for these responses. Similarly for the regulation of neutrophil migration, β-glucans must be presented to CR3 in the context of other components such as fibronectin [186].

As alluded to earlier, investigations of β-glucans and CR3 have primarily focused on the antitumorogenic potential of their interaction. Binding of β-glucans to CR3 triggers signalling *via* Syk and phosphatidylinositol 3-kinase that results in the priming of leukocytes for CR3-mediated cytotoxicity of iC3b-coated tumor cells [189, 190]. This approach has been successfully applied in combination with monoclonal antibody targeting and *in vivo* studies have shown CR3-dependent enhanced tumor killing and survival [49]. A more in depth discussion of this receptor and topic is included in the chapter which focuses on β-glucans and cancer.

LACTOSYLCERAMIDE

Lactosylceramide (LacCer; CDw17; Galβ4Glcβ1Cer) is a member of the glycosphingolipid (GSL) family. These are membrane proteins consisting of a hydrophobic ceramide moiety and an extracellular hydrophilic oligosaccharide region. The ceramide component is responsible for anchoring the protein into the outer layer of plasma membranes. Glycosphingolipids function in normal cell adhesion, migration, and proliferation and also have roles in pathologic conditions such as tumorigenesis and atherosclerosis [191, 192]. They interact with other membrane components such as intergrins, cholesterol and non-receptor cytoplasmic protein kinases to form clusters which are commonly called lipid rafts [193].

Lactosylceramide is present in the membranes of many cell types and is particularly abundant in epithelial cells and neutrophils. It has been shown to bind specifically to a variety of microbes and proposed to function as an adhesion receptor between these pathogens and host cells [194]. The specificity of lactosylceramide for β(1-3)-glucans was initially demonstrated following a biochemical screen of leukocytes with a radiolabelled high molecular weight β-glucan [195]. A limited number of *in vitro* studies have since shown that the interaction between β-glucans and lactosylceramide mediates a number of cellular responses. For example, β-glucans derived from the cell wall of *P. carinii* stimulated cytokine production following PKC dependent activation of NF-κB. [196, 197]. It has also been demonstrated that β-glucan binding to lactosylceramide enhanced the respiratory burst and microbicidal activity of leukocytes [198]. An independent study using an anti-LacCer antibody as a stimulus demonstrated that binding to the receptor induced the activation of the Src kinase Lyn which was associated in the lipid raft, leading to superoxide generation *via* a pathway that was dependent on phosphotidylinositol-3 kinase, p28 MAPK, and PKC [199]. It has also been recently shown that β-glucan activation of CR3 causes its translocation into Lyn-coupled, lactosylceramide-enriched lipid rafts allowing signalling and CR3-mediated phagocytosis to occur [200]. Finally, an investigation using a specific branched β-glucan derived from *C. albicans* showed binding to lactosylceramide and induction of neutrophils migration through the activation of Lyn and phosphatidylinositol-3 kinase [201].

LANGERIN (CD207)

Langerin (CD207) is a type II C-type lectin receptor consisting of a single extracellular CTLD connected to a neck and a short cytoplasmic tail that contains a proline rich motif [202]. It exists as an oligomer, forming trimers stabilised by a coiled-coil of α-helices in the neck region [203]. Langerin expression was originally thought to be unique to langerhans cells, a type of dendritic cell that is localised in the epidermal layer of the skin [204]. Recent studies have however suggested that it can also be expressed by a population of dermal dendritic cells [204]. Furthermore langerin⁺ dendritic cells have also been identified in the human gut, lung epithelium and kidney [204].

The extracellular region of langerin contains the residues shown to be necessary for Ca^{2+}-dependent binding of carbohydrates to C-type lectins and indeed it has been shown to bind to various sugars [203]. It has also been shown to bind to a number of pathogens including HIV-1, *Mycobacterium leprae*, a variety of *Candida* and *Saccharomyces* species and *Malassezia furfur* [205-209]. Crosslinking of langerin induces the formation of Birbeck granules, unique organelles of langerhans cells that are poorly understood but proposed to function in antigen presentation [207].

Langerin has very recently been shown to recognise β-glucans and it was suggested that its recognition of fungi is mediated *via* β-glucan binding [205]. Whether langerin can capture fungi as it does with HIV-1 remains unclear and although langerhans cells can phagocytose fungi to a certain extent, the process is not very efficient [205]. This is a very recent study and future work will establish whether langerin contributes significantly to antifungal immunity.

SCAVENGER RECEPTORS

The Scavenger Receptor family are a large heterogeneous group of structurally diverse proteins involved in uptake of modified low density lipoproteins, selected polyanionic ligands and a variety of microbes [210]. Scavenger receptors are expressed by myeloid cells and certain endothelial cells and have functions relating to immunity and homeostasis [210]. Several studies had implicated scavenger receptors in β-glucan recognition [211-213], but until very recently only a *Drosophila melanogaster* scavenger receptor, dSR-CI, had been shown to specifically bind these carbohydrates [214]. However in the past two years, four other scavenger receptors have been specifically implicated in β-glucan recognition.

CD5

CD5 is a scavenger receptor composed of three extracellular scavenger receptor cysteine rich repeats, a transmembrane region and a cytoplasmic tail containing an inhibitory signalling motif. It is expressed on T cells and a small subset of mature B cells where it associates with antigen receptors. It is thought that CD5 plays an inhibitory role in B-cell receptor signalling as evidenced by its association with the tyrosine phosphatase SHP-1 in B cells [215]. A soluble isoform of human CD5 has recently been shown to bind β-glucan and a number of yeast cell walls [216]. Furthermore, cell lines transfected with CD5 produced IL-8 in response to zymosan stimulation suggesting a pro-inflammatory role for CD5 [216]. In addition, sustained phosphorylation of the MAPKs, Erk1/2 and MEK was observed in CD5-expressing cells following zymosan stimulation [216].

Collectin Placenta 1 (CL-P1)

CL-P1, also known as scavenger receptor with C-type lectins (SRCL) is an endothelial type II glycoprotein containing a cytoplasmic tail, a single transmembrane region and an extracellular region consisting of a coiled coil domain, a collagen domain and a CTLD [217-219]. Structurally it is closely related to macrophage scavenger receptor class A and it has been demonstrated to occur clustered in trimers [220]. CL-P1 has been shown to bind oxidised LDL, several carbohydrates, various bacteria and yeast, and carcinoma associated antigens [217-222]. It has also been shown that CL-P1 binding of bacteria and yeast mediates microbial uptake [217-219]. CL-P1 has recently been implicated in binding fibrillar β-amyloid protein, suggesting that it may be involved in the clearance and/or pathogenesis of Alzheimers disease [223]. Although not directly shown, a recent study demonstrating that CL-P1 expressing cells could bind and phagocytose zymosan suggested a function for this receptor in β-glucan recognition [224].

CD36

CD36 is a highly glycosylated member of the class B scavenger receptors and consists of a single polypeptide chain anchored in cell membranes. It is expressed by many cell types and interacts with a large variety of ligands including collagen, thrombospondin, oxidised LDL, fatty acids and microbial products [225]. CD36 has recently been shown to bind *C. neoformans* and *C. albicans* in a β-glucan dependent manner [226]. Furthermore macrophages from CD36 deficient mice showed decreased binding, internalisation and cytokine production in response to these fungi [226]. *In vivo* studies also demonstrated that CD36 expression is required for *C. neoformans* induced cytokine and chemokine production and plays a crucial role in the host response to this organism [226].

SCARF1

SCARF1 also known as SREC (Scavenger Receptor from Endothelial Cells), is a type F scavenger receptor which has an unusually long intracellular tail [227]. SCARF1 expression has been demonstrated in endothelial cells and macrophages and it was shown to mediate the endocytosis of chemically modified lipoproteins and other polyanionic substrates as well as the chaperones gp96 and calreticulin [227-229]. During the same investigation that identified the β-glucan binding properties of CD36, SCARF1 was also implicated in β-glucan recognition and the innate response to *C. neoformans* and *C. albicans*. Furthermore, it was found that coexpression of SCARF1 and TLR2 synergized for maximal responsiveness to *C. neoformans*. It is possible that like Dectin-1, SCARF1 may also collaborate with other TLRs to enhance immune responses.

CONCLUSIONS

Host recognition of β-glucans appears to be of central importance during fungal infections. The identification and characterisation of specific β-glucan receptors has provided important insights into the mechanisms underlying the activities of β-glucans and anti-fungal immunity, but also innate immunity more generally. It is likely that other receptors for these carbohydrates will emerge in the future. Continued investigations are required to enhance our understanding of these mechanisms in order to develop strategies to improve immune responses to fungal infection and to exploit the immunomodulating capacity of β-glucans as therapeutics.

ACKNOWLEDGEMENTS

We thank the Biotechnology and Biological Sciences Research Council (BBSRC) and Welcome Trust for funding.

REFERENCES

[1]   Iwasaki A, Medzhitov R. Regulation of adaptive immunity by the innate immune system. Science 2010; 327: 291-5.
[2]   Janeway CA, Jr. The immune system evolved to discriminate infectious nonself from noninfectious self. Immunol Today 1992; 13: 11-6.
[3]   Janeway CA, Jr., Medzhitov R. Innate immune recognition. Annl Rev Immunol 2002; 20: 197-216.
[4]   Michelsen KS, Arditi M. Toll-like receptors and innate immunity in gut homeostasis and pathology. Curr Opin Hematol 2007; 14: 48-54.
[5]   Yamasaki S, Ishikawa E, Sakuma M, *et al.* Mincle is an ITAM-coupled activating receptor that senses damaged cells. Nat Immunol 2008; 9: 1179-88.
[6]   Sancho D, Joffre OP, Keller AM, *et al.* Identification of a dendritic cell receptor that couples sensing of necrosis to immunity. Nature 2009; 458: 899-903.
[7]   Pasare C, Medzhitov R. Toll-like receptors: linking innate and adaptive immunity. Adv Exp Med Biol 2005; 560: 11-8.
[8]   Kawai T, Akira S. The role of pattern-recognition receptors in innate immunity: update on Toll-like receptors. Nat Immunol 2010; 11: 373-84.
[9]   Moore CB, Ting JP. Regulation of mitochondrial antiviral signaling pathways. Immunity 2008; 28: 735-9.
[10]   Kawai T, Akira S. Innate immune recognition of viral infection. Nat Immunol 2006; 7: 131-7.
[11]   Franchi L, Warner N, Viani K, *et al.* Function of Nod-like receptors in microbial recognition and host defense. Immunol Rev 2009; 227: 106-28.
[12]   Huysamen C, Brown GD. The fungal pattern recognition receptor, Dectin-1, and the associated cluster of C-type lectin-like receptors. FEMS Microbiol Lett 2009; 290: 121-8.

[13]   Willment JA, Brown GD. C-type lectin receptors in antifungal immunity. Trends Microbiol 2008; 16: 27-32.

[14]   Robinson MJ, Sancho D, Slack EC, *et al.* Myeloid C-type lectins in innate immunity. Nat Immunol 2006; 7: 1258-65.

[15]   Palm NW, Medzhitov R. Pattern recognition receptors and control of adaptive immunity. Immunol Rev 2009; 227: 221-33.

[16]   Park SJ, Mehrad B. Innate immunity to *Aspergillus* species. Clin Microbiol Rev 2009; 22: 535-51.

[17]   Zelante T, Montagnoli C, Bozza S, *et al.* Receptors and pathways in innate antifungal immunity: the implication for tolerance and immunity to fungi. Adv Exp Med Biol 2007; 590: 209-21.

[18]   Klis FM, Mol P, Hellingwerf K, *et al.* Dynamics of cell wall structure in *Saccharomyces cerevisiae.* FEMS Microbiol Rev 2002; 26: 239-56.

[19]   Kollar R, Petrakova E, Ashwell G, *et al.* Architecture of the yeast cell wall. The linkage between chitin and beta(1-->3)-glucan. J Biol Chem 1995; 270: 1170-8.

[20]   Kollar R, Reinhold BB, Petrakova E, *et al.* Architecture of the yeast cell wall. Beta(1-->6)-glucan interconnects mannoprotein, beta(1-->)3-glucan, and chitin. J Biol Chem 1997; 272: 17762-75.

[21]   Lipke PN, Ovalle R. Cell wall architecture in yeast: new structure and new challenges. J Bacteriol 1998; 180: 3735-40.

[22]   Riggi SJ, Di Luzio NR. Identification of a reticuloendothelial stimulating agent in zymosan. Am J Physiol 1961; 200: 297-300.

[23]   Okazaki M, Adachi Y, Ohno N, *et al.* Structure-activity relationship of (1-->3)-beta-D-glucans in the induction of cytokine production from macrophages, *in vitro.* Biol Pharm Bull 1995; 18: 1320-7.

[24]   Kulicke WM, Lettau AI, Thielking H. Correlation between immunological activity, molar mass, and molecular structure of different (1-->3)-beta-D-glucans. Carbohydr Res 1997; 297: 135-43.

[25]   Ohno N, Asada N, Adachi Y, *et al..* Enhancement of LPS triggered TNF-alpha (tumor necrosis factor-alpha) production by (1-->3)-beta-D-glucans in mice. Biol Pharm Bull 1995; 18: 126-33.

[26]   Yoshioka Y, Uehara N, Saito H. Conformation-dependent change in antitumor activity of linear and branched (1---->3)-beta-D-glucans on the basis of conformational elucidation by carbon-13 nuclear magnetic resonance spectroscopy. Chem Pharm Bull (Tokyo) 1992; 40: 1221-6.

[27]   Mueller A, Raptis J, Rice PJ, *et al.* The influence of glucan polymer structure and solution conformation on binding to (1-->3)-beta-D-glucan receptors in a human monocyte-like cell line. Glycobiology 2000; 10: 339-46.

[28]   Adams EL, Rice PJ, Graves B, *et al.* Differential high-affinity interaction of dectin-1 with natural or synthetic glucans is dependent upon primary structure and is influenced by polymer chain length and side-chain branching. J Pharmacol Exp Therap 2008; 325: 115-23.

[29]   Brown GD, Herre J, Williams DL, *et al.* Dectin-1 mediates the biological effects of beta-glucans. J Exp Med 2003; 197: 1119-24.

[30]   Czop JK, Austen KF. A beta-glucan inhibitable receptor on human monocytes: its identity with the phagocytic receptor for particulate activators of the alternative complement pathway. J Immunol 1985; 134: 2588-93.

[31]   Dillon S, Agrawal S, Banerjee K, *et al.* Yeast zymosan, a stimulus for TLR2 and dectin-1, induces regulatory antigen-presenting cells and immunological tolerance. J Clin Invest 2006; 116: 916-28.

[32]   Hida TH, Ishibashi K, Miura NN, *et al.* Cytokine induction by a linear 1,3-glucan, curdlan-oligo, in mouse leukocytes *in vitro.* Inflamm Res 2009; 58: 9-14.

[33]   Li C, Ha T, Kelley J, *et al.* Modulating Toll-like receptor mediated signaling by (1-->3)-beta-D-glucan rapidly induces cardioprotection. Cardiovas Res 2004; 61: 538-47.

[34]   Adams DS, Pero SC, Petro JB, *et al.* PGG-Glucan activates NF-kappaB-like and NF-IL-6-like transcription factor complexes in a murine monocytic cell line. J Leukoc Biol 1997; 62: 865-73.

[35]   Battle J, Ha T, Li C, *et al.* Ligand binding to the (1 --> 3)-beta-D-glucan receptor stimulates NFkappaB activation, but not apoptosis in U937 cells. Biochem Biophys Res Commun 1998; 249: 499-504.

[36]   Williams DL, Li C, Ha T, *et al.* Modulation of the phosphoinositide 3-kinase pathway alters innate resistance to polymicrobial sepsis. J Immunol 2004; 172: 449-56.

[37]   Brown GD, Gordon S. Immune recognition. A new receptor for beta-glucans. Nature 2001; 413: 36-7.

[38]   Ross GD, Cain JA, Myones BL, *et al.* Specificity of membrane complement receptor type three (CR3) for beta-glucans. Complement 1987; 4: 61-74.

[39]   Tzianabos AO. Polysaccharide immuno-modulators as therapeutic agents: structural aspects and biologic function. Clin Microbiol Rev 2000; 13: 523-33.

[40]   Chen J, Seviour R. Medicinal importance of fungal beta-(1-->3), (1-->6)-glucans. Mycol Res 2007; 111: 635-52.

[41]   Chae BJ, Lohakare JD, Moon WK, *et al.* Effects of supplementation of beta-glucan on the growth performance and immunity in broilers. Res Vet Sci 2006; 80: 291-8.

[42] Cook MT, Hayball PJ, Hutchinson W, *et al.* Administration of a commercial immunostimulant preparation, EcoActiva as a feed supplement enhances macrophage respiratory burst and the growth rate of snapper (*Pagrus auratus*, Sparidae (Bloch and Schneider)) in winter. Fish Shellfish Immunol 2003; 14: 333-45.

[43] Van Hai N, Buller N, Fotedar R. The use of customised probiotics in the cultivation of western king prawns (*Penaeus latisulcatus* Kishinouye, 1896). Fish Shellfish Immunol 2009; 27: 100-4.

[44] Stuyven E, Cox E, Vancaeneghem S, *et al.* Effect of beta-glucans on an ETEC infection in piglets. Vet Immunol Immunopathol 2009; 128: 60-6.

[45] Dellinger EP, Babineau TJ, Bleicher P, *et al.* Effect of PGG-glucan on the rate of serious postoperative infection or death observed after high-risk gastrointestinal operations. Betafectin Gastrointestinal Study Group. Arch Surg 1999; 134: 977-83.

[46] Bradner WT, Clarke DA, Stock CC. Stimulation of host defense against experimental cancer. I. Zymosan and sarcoma 180 in mice. Cancer Res 1958; 18: 347-51.

[47] Yan J, Allendorf DJ, Brandley B. Yeast whole glucan particle (WGP) beta-glucan in conjunction with antitumour monoclonal antibodies to treat cancer. Expert Opin Biol Ther 2005; 5: 691-702.

[48] Liu J, Gunn L, Hansen R, Yan J. Combined yeast-derived beta-glucan with anti-tumor monoclonal antibody for cancer immunotherapy. Exp Mol Pathol 2009; 86: 208-14.

[49] Gelderman KA, Tomlinson S, Ross GD, *et al.* Complement function in mAb-mediated cancer immunotherapy. Trends Immunol 2004; 25: 158-64.

[50] Ross GD, Vetvicka V, Yan J, *et al.* Therapeutic intervention with complement and beta-glucan in cancer. Immunopharmacology 1999; 42: 61-74.

[51] Cheung NK, Modak S, Vickers A, *et al.* Orally administered beta-glucans enhance anti-tumor effects of monoclonal antibodies. Cancer Immunol Immunother 2002; 51: 557-64.

[52] Cheung NK, Modak S. Oral (1-->3),(1-->4)-beta-D-glucan synergizes with antiganglioside GD2 monoclonal antibody 3F8 in the therapy of neuroblastoma. Clin Cancer Res 2002; 8: 1217-23.

[53] Hong F, Hansen RD, Yan J, *et al.* Beta-glucan functions as an adjuvant for monoclonal antibody immunotherapy by recruiting tumoricidal granulocytes as killer cells. Cancer Res 2003; 63: 9023-31.

[54] Hong F, Yan J, Baran JT, *et al.* Mechanism by which orally administered beta-1,3-glucans enhance the tumoricidal activity of antitumor monoclonal antibodies in murine tumor models. J Immunol 2004; 173: 797-806.

[55] Modak S, Koehne G, Vickers A, *et al.*. Rituximab therapy of lymphoma is enhanced by orally administered (1-->3),(1-->4)-D-beta-glucan. Leuk Res 2005; 29: 679-83.

[56] Salvador C, Li B, Hansen R, *et al.* Yeast-derived beta-glucan augments the therapeutic efficacy mediated by anti-vascular endothelial growth factor monoclonal antibody in human carcinoma xenograft models. Clin Cancer Res 2008; 14: 1239-47.

[57] Yan J, Vetvicka V, Xia Y, *et al.* Beta-glucan, a "specific" biologic response modifier that uses antibodies to target tumors for cytotoxic recognition by leukocyte complement receptor type 3 (CD11b/CD18). J Immunol 1999; 163: 3045-52.

[58] Bell S, Goldman VM, Bistrian BR, *et al.* Effect of beta-glucan from oats and yeast on serum lipids. Crit Rev Food Sci Nutr 1999; 39: 189-202.

[59] Theuwissen E, Mensink RP. Water-soluble dietary fibers and cardiovascular disease. Physiol Behav 2008; 94: 285-92.

[60] Maki KC, Galant R, Samuel P, *et al.* Effects of consuming foods containing oat beta-glucan on blood pressure, carbohydrate metabolism and biomarkers of oxidative stress in men and women with elevated blood pressure. Eur J Clin Nutr 2007; 61: 786-95.

[61] Chen J, Raymond K. Beta-glucans in the treatment of diabetes and associated cardiovascular risks. Vasc Health Risk Manag 2008; 4: 1265-72.

[62] Battilana P, Ornstein K, Minehira K, *et al.* Mechanisms of action of beta-glucan in postprandial glucose metabolism in healthy men. Eur J Clin Nutr 2001; 55: 327-33.

[63] Cook JA, Dougherty WJ, Holt TM. Enhanced sensitivity to endotoxin induced by the RE stimulant, glucan. Circ Shock 1980; 7: 225-38.

[64] Bowers GJ, Patchen ML, MacVittie TJ, *et al.* A comparative evaluation of particulate and soluble glucan in an endotoxin model. Int J Immunopharmacol 1986; 8: 313-21.

[65] Pretus HA, Ensley HE, McNamee RB, *et al.* Isolation, physicochemical characterization and preclinical efficacy evaluation of soluble scleroglucan. J Pharmacol Exp Therapeut 1991; 257: 500-10.

[66] Tsoni SV, Brown GD. beta-Glucans and dectin-1. Ann New York Acad Sci 2008; 1143: 45-60.

[67] Yoshioka S, Ohno N, Miura T, *et al.* Immunotoxicity of soluble beta-glucans induced by indomethacin treatment. FEMS Immunol Med Microbiol 1998; 21: 171-9.

[68]   Takahashi H, Ohno N, Adachi Y, *et al.* Association of immunological disorders in lethal side effect of NSAIDs on beta-glucan-administered mice. FEMS Immunol Med Microbiol 2001; 31: 1-14.

[69]   Nameda S, Miura NN, Adachi Y, *et al.* Lincomycin protects mice from septic shock in beta-glucan-indomethacin model. Biol Pharmaceut Bull 2007; 30: 2312-6.

[70]   Nameda S, Miura NN, Adachi Y, *et al.* Antibiotics protect against septic shock in mice administered beta-glucan and indomethacin. Microbiol Immunol 2007; 51: 851-9.

[71]   Vetvicka V, Vetvickova J. Beta-glucan-indomethacin combination produces no lethal effects. Biomed Pap Med Fac Univ Palacky Olomouc Czech Repub 2009; 153: 111-6.

[72]   Rylander R, Lin RH. (1-->3)-beta-D-glucan - relationship to indoor air-related symptoms, allergy and asthma. Toxicology 2000; 152: 47-52.

[73]   Douwes J. (1-->3)-Beta-D-glucans and respiratory health: a review of the scientific evidence. Indoor Air 2005; 15: 160-9.

[74]   Yoshitomi H, Sakaguchi N, Kobayashi K, *et al.* A role for fungal {beta}-glucans and their receptor Dectin-1 in the induction of autoimmune arthritis in genetically susceptible mice. J Exp Med 2005; 201: 949-60.

[75]   Hida S, Miura NN, Adachi Y, *et al.* Cell wall beta-glucan derived from *Candida albicans* acts as a trigger for autoimmune arthritis in SKG mice. Biol Pharm Bull 2007; 30: 1589-92.

[76]   Czop JK. The role of beta-glucan receptors on blood and tissue leukocytes in phagocytosis and metabolic activation. Pathol Immunopathol Res 1986; 5: 286-96.

[77]   Rogers NC, Slack EC, Edwards AD, *et al.* Syk-dependent cytokine induction by Dectin-1 reveals a novel pattern recognition pathway for C type lectins. Immunity 2005; 22: 507-17.

[78]   Willment JA, Marshall AS, Reid DM, *et al.* The human beta-glucan receptor is widely expressed and functionally equivalent to murine Dectin-1 on primary cells. Eur J Immunol 2005; 35: 1539-47.

[79]   Brown GD, Taylor PR, Reid DM, *et al.* Dectin-1 is a major beta-glucan receptor on macrophages. J Exp Med 2002; 196: 407-12.

[80]   Gantner BN, Simmons RM, Canavera SJ, *et al.* Collaborative induction of inflammatory responses by dectin-1 and Toll-like receptor 2. J Exp Med 2003; 197: 1107-17.

[81]   Drickamer K. C-type lectin-like domains. Curr Opin Struct Biol 1999; 9: 585-90.

[82]   Weis WI, Taylor ME, Drickamer K. The C-type lectin superfamily in the immune system. Immunol Rev 1998; 163: 19-34.

[83]   Brown GD. Dectin-1: a signalling non-TLR pattern-recognition receptor. Nat Rev Immunol 2006; 6: 33-43.

[84]   Sobanov Y, Bernreiter A, Derdak S, *et al.* A novel cluster of lectin-like receptor genes expressed in monocytic, dendritic and endothelial cells maps close to the NK receptor genes in the human NK gene complex. Eur J Immunol 2001; 31: 3493-503.

[85]   Marshall AS, Willment JA, Lin HH, *et al.* Identification and characterization of a novel human myeloid inhibitory C-type lectin-like receptor (MICL) that is predominantly expressed on granulocytes and monocytes. J Biol Chem 2004; 279: 14792-802.

[86]   Heinsbroek SE, Taylor PR, Rosas M, *et al.* Expression of functionally different dectin-1 isoforms by murine macrophages. J Immunol 2006; 176: 5513-8.

[87]   Adachi Y, Ishii T, Ikeda Y, *et al.* Characterization of beta-glucan recognition site on C-type lectin, dectin 1. Infect Immun 2004; 72: 4159-71.

[88]   Willment JA, Gordon S, Brown GD. Characterization of the human beta -glucan receptor and its alternatively spliced isoforms. J Biol Chem 2001; 276: 43818-23.

[89]   Yokota K, Takashima A, Bergstresser PR, *et al.* Identification of a human homologue of the dendritic cell-associated C-type lectin-1, dectin-1. Gene 2001; 272: 51-60.

[90]   Xie J, Sun M, Guo L, *et al.* Human Dectin-1 isoform E is a cytoplasmic protein and interacts with RanBPM. Biochem Biophys Res Commun 2006; 347: 1067-73.

[91]   Taylor PR, Brown GD, Reid DM, *et al.* The beta-glucan receptor, dectin-1, is predominantly expressed on the surface of cells of the monocyte/macrophage and neutrophil lineages. J Immunol 2002; 169: 3876-82.

[92]   Olynych TJ, Jakeman DL, Marshall JS. Fungal zymosan induces leukotriene production by human mast cells through a dectin-1-dependent mechanism. J Allergy Clin Immunol 2006; 118: 837-43.

[93]   Shah VB, Huang Y, Keshwara R, *et al.* Beta-glucan activates microglia without inducing cytokine production in Dectin-1-dependent manner. J Immunol 2008; 180: 2777-85.

[94]   Reid DM, Montoya M, Taylor PR, *et al.* Expression of the beta-glucan receptor, Dectin-1, on murine leukocytes in situ correlates with its function in pathogen recognition and reveals potential roles in leukocyte interactions. J Leukoc Biol 2004; 76: 86-94.

[95]   Kato Y, Adachi Y, Ohno N. Contribution of N-linked oligosaccharides to the expression and functions of beta-glucan receptor, Dectin-1. Biol Pharmaceut Bull 2006; 29: 1580-6.

[96]   Willment JA, Lin HH, Reid DM, *et al.* Dectin-1 expression and function are enhanced on alternatively activated and GM-CSF-treated macrophages and are negatively regulated by IL-10, dexamethasone, and lipopolysaccharide. J Immunol 2003; 171: 4569-73.

[97]   Ozment-Skelton TR, Goldman MP, Gordon S, *et al.* Prolonged reduction of leukocyte membrane-associated Dectin-1 levels following beta-glucan administration. J Pharmacol Exp Therapeut 2006; 318: 540-6.

[98]   Ozment-Skelton TR, deFluiter EA, Ha T, *et al.* Leukocyte Dectin-1 expression is differentially regulated in fungal versus polymicrobial sepsis. Crit Care Med 2009; 37: 1038-45.

[99]   Ariizumi K, Shen GL, Shikano S, *et al.* Identification of a novel, dendritic cell-associated molecule, dectin-1, by subtractive cDNA cloning. J Biol Chem 2000; 275: 20157-67.

[100] Grunebach F, Weck MM, Reichert J, *et al.* Molecular and functional characterization of human Dectin-1. Exp Hematol 2002; 30: 1309-15.

[101] Palma AS, Feizi T, Zhang Y, *et al.* Ligands for the beta-glucan receptor, Dectin-1, assigned using "designer" microarrays of oligosaccharide probes (neoglycolipids) generated from glucan polysaccharides. J Biol Chem 2006; 281: 5771-9.

[102] Brown J, O'Callaghan CA, Marshall AS, *et al.* Structure of the fungal beta-glucan-binding immune receptor dectin-1: implications for function. Protein Sci 2007; 16: 1042-52.

[103] Gersuk GM, Underhill DM, Zhu L, *et al.* Dectin-1 and TLRs permit macrophages to distinguish between different Aspergillus fumigatus cellular states. J Immunol 2006; 176: 3717-24.

[104] Saijo S, Fujikado N, Furuta T, *et al.* Dectin-1 is required for host defense against *Pneumocystis carinii* but not against *Candida albicans*. Nat Immunol 2007; 8: 39-46.

[105] Steele C, Marrero L, Swain S, *et al.* Alveolar macrophage-mediated killing of *Pneumocystis carinii* f. sp. *muris* involves molecular recognition by the Dectin-1 beta-glucan receptor. J Exp Med 2003; 198: 1677-88.

[106] Steele C, Rapaka RR, Metz A, *et al.* The beta-glucan receptor dectin-1 recognizes specific morphologies of *Aspergillus fumigatus*. PLoS Pathog 2005; 1: e42.

[107] Taylor PR, Tsoni SV, Willment JA, *et al.* Dectin-1 is required for beta-glucan recognition and control of fungal infection. Nat Immunol 2007; 8: 31-8.

[108] Viriyakosol S, Fierer J, Brown GD, *et al.* Innate immunity to the pathogenic fungus *Coccidioides posadasii* is dependent on Toll-like receptor 2 and Dectin-1. Infect Immun 2005; 73: 1553-60.

[109] Gantner BN, Simmons RM, Underhill DM. Dectin-1 mediates macrophage recognition of *Candida albicans* yeast but not filaments. Embo J 2005; 24: 1277-86.

[110] Nakamura K, Miyazato A, Koguchi Y, *et al.* Toll-like receptor 2 (TLR2) and dectin-1 contribute to the production of IL-12p40 by bone marrow-derived dendritic cells infected with P*enicillium marneffei*. Microb Infect / Institut Pasteur 2008; 10: 1223-7.

[111] Rothfuchs AG, Bafica A, Feng CG, *et al.* Dectin-1 interaction with *Mycobacterium tuberculosis* leads to enhanced IL-12p40 production by splenic dendritic cells. J Immunol 2007; 179: 3463-71.

[112] Shin DM, Yang CS, Yuk JM, *et al. Mycobacterium abscessus* activates the macrophage innate immune response *via* a physical and functional interaction between TLR2 and dectin-1. Cell Microbiol 2008; 10: 1608-21.

[113] Yadav M, Schorey JS. The beta-glucan receptor dectin-1 functions together with TLR2 to mediate macrophage activation by mycobacteria. Blood 2006; 108: 3168-75.

[114] Lee HM, Shin DM, Choi DK, *et al.* Innate immune responses to *Mycobacterium ulcerans via* toll-like receptors and dectin-1 in human keratinocytes. Cell Microbiol 2009; 11: 678-92.

[115] Weck MM, Appel S, Werth D, *et al.* Dectin-1 is involved in uptake and cross-presentation of cellular antigens. Blood 2008; 111: 4264-72.

[116] LeibundGut-Landmann S, Gross O, Robinson MJ, *et al.* Syk- and CARD9-dependent coupling of innate immunity to the induction of T helper cells that produce interleukin 17. Nat Immunol 2007; 8: 630-8.

[117] Suram S, Brown GD, Ghosh M, *et al.* Regulation of cytosolic phospholipase A2 activation and cyclooxygenase 2 expression in macrophages by the beta-glucan receptor. J Biol Chem 2006; 281: 5506-14.

[118] Valera I, Fernandez N, Trinidad AG, *et al.* Costimulation of dectin-1 and DC-SIGN triggers the arachidonic acid cascade in human monocyte-derived dendritic cells. J Immunol 2008; 180: 5727-36.

[119] Olsson S, Sundler R. The macrophage beta-glucan receptor mediates arachidonate release induced by zymosan: Essential role for Src family kinases. Mol Immunol 2007; 44:1509-15.

[120] Herre J, Marshall AS, Caron E, *et al.* Dectin-1 uses novel mechanisms for yeast phagocytosis in macrophages. Blood 2004; 104: 4038-45.

[121] Underhill DM, Rossnagle E, Lowell CA, *et al.* Dectin-1 activates Syk tyrosine kinase in a dynamic subset of macrophages for reactive oxygen production. Blood 2005; 106: 2543-50.

[122] Zenaro E, Donini M, Dusi S. Induction of Th1/Th17 immune response by *Mycobacterium tuberculosis*: role of dectin-1, mannose receptor, and DC-SIGN. J Leukoc Biol 2009; 86: 1393-1401.

[123] Osorio F, LeibundGut-Landmann S, Lochner M, *et al.* DC activated *via* dectin-1 convert Treg into IL-17 producers. Eur J Immunol 2008; 38: 3274-81.

[124] Karumuthil-Melethil S, Perez N, Li R, *et al.* Induction of innate immune response through TLR2 and dectin 1 prevents type 1 diabetes. J Immunol 2008; 181: 8323-34.

[125] Leibundgut-Landmann S, Osorio F, Brown GD, Reis e Sousa C. Stimulation of dendritic cells *via* the dectin-1/Syk pathway allows priming of cytotoxic T-cell responses. Blood 2008; 112: 4971-80.

[126] Carter RW, Thompson C, Reid DM, *et al.* Preferential induction of CD4+ T cell responses through *in vivo* targeting of antigen to dendritic cell-associated C-type lectin-1. J Immunol 2006; 177: 2276-84.

[127] Dennehy KM, Ferwerda G, Faro-Trindade I, *et al.* Syk kinase is required for collaborative cytokine production induced through Dectin-1 and Toll-like receptors. Eur J Immunol 2008; 38: 500-6.

[128] Goodridge HS, Shimada T, Wolf AJ, *et al.* Differential use of CARD9 by dectin-1 in macrophages and dendritic cells. J Immunol 2009; 182: 1146-54.

[129] Rosas M, Liddiard K, Kimberg M, *et al.* The induction of inflammation by dectin-1 *in vivo* is dependent on myeloid cell programming and the progression of phagocytosis. J Immunol 2008; 181: 3549-57.

[130] Isakov N. Immunoreceptor tyrosine-based activation motif (ITAM), a unique module linking antigen and Fc receptors to their signaling cascades. J Leukoc Biol 1997; 61: 6-16.

[131] Underhill DM, Goodridge HS. The many faces of ITAMs. Trends Immunol 2007; 28: 66-73.

[132] Barrow AD, Trowsdale J. You say ITAM and I say ITIM, let's call the whole thing off: the ambiguity of immunoreceptor signalling. Eur J Immunol 2006; 36: 1646-53.

[133] Fuller GL, Williams JA, Tomlinson MG, *et al.* The C-type lectin receptors CLEC-2 and Dectin-1, but not DC-SIGN, signal *via* a novel YXXL-dependent signaling cascade. J Biol Chem 2007; 282: 12397-409.

[134] Huysamen C, Willment JA, Dennehy KM, *et al.* CLEC9A is a novel activation C-type lectin-like receptor expressed on BDCA3+ dendritic cells and a subset of monocytes. J Biol Chem 2008; 283: 16693-701.

[135] Goodridge HS, Wolf AJ, Underhill DM. Beta-glucan recognition by the innate immune system. Immunol Rev 2009; 230: 38-50.

[136] Gross O, Gewies A, Finger K, *et al.* Card9 controls a non-TLR signalling pathway for innate anti-fungal immunity. Nature 2006; 442: 651-6.

[137] Tassi I, Cella M, Castro I, *et al.* Requirement of phospholipase C-gamma2 (PLCgamma2) for Dectin-1-induced antigen presentation and induction of TH1/TH17 polarization. Eur J Immunol 2009; 39: 1369-78.

[138] Xu S, Huo J, Lee KG, *et al.* Phospholipase Cgamma2 is critical for Dectin-1-mediated Ca2+ flux and cytokine production in dendritic cells. J Biol Chem 2009; 284: 7038-46.

[139] Dennehy KM, Brown GD. The role of the beta-glucan receptor Dectin-1 in control of fungal infection. J Leukoc Biol 2007; 82: 253-8.

[140] Gringhuis SI, den Dunnen J, Litjens M, *et al.* Dectin-1 directs T helper cell differentiation by controlling noncanonical NF-kappaB activation through Raf-1 and Syk. Nat Immunol 2009; 10: 203-13.

[141] Goodridge HS, Simmons RM, Underhill DM. Dectin-1 stimulation by *Candida albicans* yeast or zymosan triggers NFAT activation in macrophages and dendritic cells. J Immunol 2007; 178: 3107-15.

[142] Slack EC, Robinson MJ, Hernanz-Falcon P, *et al.* Syk-dependent ERK activation regulates IL-2 and IL-10 production by DC stimulated with zymosan. Eur J Immunol 2007; 37: 1600-12

[143] Trinchieri G, Sher A. Cooperation of Toll-like receptor signals in innate immune defence. Nat Rev Immunol 2007; 7: 179-90.

[144] Reid DM, Gow NA, Brown GD. Pattern recognition: recent insights from Dectin-1. Curr Opin Immunol 2009; 21: 30-7.

[145] Ferwerda G, Meyer-Wentrup F, Kullberg BJ, *et al.* Dectin-1 synergizes with TLR2 and TLR4 for cytokine production in human primary monocytes and macrophages. Cell Microbiol 2008; 10: 2058-66.

[146] Gerosa F, Baldani-Guerra B, Lyakh LA, *et al.* Differential regulation of interleukin 12 and interleukin 23 production in human dendritic cells. J Exp Med 2008; 205: 1447-61.

[147] Geijtenbeek TB, Gringhuis SI. Signalling through C-type lectin receptors: shaping immune responses. Nat Rev Immunol 2009; 9: 465-79.

[148] Taylor PR, Brown GD, Herre J, *et al.* The role of SIGNR1 and the beta-glucan receptor (dectin-1) in the nonopsonic recognition of yeast by specific macrophages. J Immunol 2004; 172: 1157-62.

[149] Boucheix C, Rubinstein E. Tetraspanins. Cell Mol Life Sci 2001; 58: 1189-205.

[150] Mantegazza AR, Barrio MM, Moutel S, *et al.* CD63 tetraspanin slows down cell migration and translocates to the endosomal-lysosomal-MIICs route after extracellular stimuli in human immature dendritic cells. Blood 2004; 104: 1183-90.

[151] Meyer-Wentrup F, Figdor CG, Ansems M, *et al.* Dectin-1 Interaction with Tetraspanin CD37 Inhibits IL-6 Production. J Immunol 2007; 178: 154-62.

[152] van Spriel AB, Sofi M, Gartlan KH, *et al.* The tetraspanin protein CD37 regulates IgA responses and anti-fungal immunity. PLoS Pathog 2009; 5: e1000338.

[153] Romani L, Puccetti P. Protective tolerance to fungi: the role of IL-10 and tryptophan catabolism. Trends Microbiol 2006; 14: 183-9.

[154] Netea MG, Sutmuller R, Hermann C, *et al.* Toll-like receptor 2 suppresses immunity against *Candida albicans* through induction of IL-10 and regulatory T cells. J Immunol 2004; 172: 3712-8.

[155] Romani L. Immunity to fungal infections. Nat Rev Immunol 2004; 4: 1-23.

[156] Huang W, Na L, Fidel PL, Schwarzenberger P. Requirement of interleukin-17A for systemic anti-*Candida albicans* host defense in mice. J Infect Dis 2004; 190: 624-31.

[157] Bozza S, Zelante T, Moretti S, *et al.* Lack of Toll IL-1R8 exacerbates Th17 cell responses in fungal infection. J Immunol 2008; 180: 4022-31.

[158] Zelante T, De Luca A, Bonifazi P, *et al.* IL-23 and the Th17 pathway promote inflammation and impair antifungal immune resistance. Eur J Immunol 2007; 37: 2695-706.

[159] Conti HR, Shen F, Nayyar N, *et al.* Th17 cells and IL-17 receptor signaling are essential for mucosal host defense against oral candidiasis. J Exp Med 2009; 206: 299-311.

[160] Leigh JE, McNulty KM, Fidel PL, Jr. Characterization of the immune status of CD8+ T cells in oral lesions of human immunodeficiency virus-infected persons with oropharyngeal candidiasis. Clin Vaccine Immunol 2006; 13: 678-83

[161] Lindell DM, Moore TA, McDonald RA, *et al.* Generation of antifungal effector CD8+ T cells in the absence of CD4+ T cells during *Cryptococcus neoformans* infection. J Immunol 2005; 174: 7920-8.

[162] Marquis M, Lewandowski D, Dugas V, *et al.* CD8+ T cells but not polymorphonuclear leukocytes are required to limit chronic oral carriage of *Candida albicans* in transgenic mice expressing human immunodeficiency virus type 1. Infect Immun 2006; 74: 2382-91.

[163] Wuthrich M, Filutowicz HI, Warner T, *et al.* Vaccine immunity to pathogenic fungi overcomes the requirement for CD4 help in exogenous antigen presentation to CD8+ T cells: implications for vaccine development in immune-deficient hosts. J Exp Med 2003; 197: 1405-16.

[164] Hise AG, Tomalka J, Ganesan S, *et al.* An essential role for the NLRP3 inflammasome in host defense against the human fungal pathogen *Candida albicans.* Cell Host Microbe 2009; 5: 487-97.

[165] Werner JL, Metz AE, Horn D, *et al.* Requisite role for the dectin-1 beta-glucan receptor in pulmonary defense against *Aspergillus fumigatus.* J Immunol 2009; 182: 4938-46.

[166] Nakamura K, Kinjo T, Saijo S, *et al.* Dectin-1 is not required for the host defense to *Cryptococcus neoformans.* Microbiol Immunol 2007; 51: 1115-9.

[167] Ferwerda B, Ferwerda G, Plantinga TS, *et al.* Human dectin-1 deficiency and mucocutaneous fungal infections. New England J Med 2009; 361: 1760-7.

[168] Glocker EO, Hennigs A, Nabavi M, *et al.* A homozygous CARD9 mutation in a family with susceptibility to fungal infections. New England J Med 2009; 361: 1727-35.

[169] Hohl TM, Van Epps HL, Rivera A, *et al. Aspergillus fumigatus* triggers inflammatory responses by stage-specific beta-glucan display. PLoS Pathog 2005; 1: e30.

[170] Torosantucci A, Bromuro C, Chiani P, *et al.* A novel glyco-conjugate vaccine against fungal pathogens. J Exp Med 2005; 202: 597-606.

[171] Wheeler RT, Fink GR. A drug-sensitive genetic network masks fungi from the immune system. PLoS Pathog 2006; 2: e35.

[172] Lamaris GA, Lewis RE, Chamilos G, *et al.* Caspofungin-mediated beta-glucan unmasking and enhancement of human polymorphonuclear neutrophil activity against *Aspergillus* and non-*Aspergillus* hyphae. J Infect Dis 2008; 198: 186-92.

[173] Wheeler RT, Kombe D, Agarwala SD, *et al.* Dynamic, morphotype-specific *Candida albicans* beta-glucan exposure during infection and drug treatment. PLoS Pathog 2008; 4: e1000227.

[174] Hohl TM, Feldmesser M, Perlin DS, Pamer EG. Caspofungin modulates inflammatory responses to *Aspergillus fumigatus* through stage-specific effects on fungal beta-glucan exposure. J Infect Dis 2008; 198: 176-85.

[175] Ikeda Y, Adachi Y, Ishii T, *et al.* Blocking effect of anti-Dectin-1 antibodies on the anti-tumor activity of 1,3-beta-glucan and the binding of Dectin-1 to 1,3-beta-glucan. Biol Pharmaceut Bull 2007; 30: 1384-9.

[176]   Williams DL, McNamee RB, Jones EL, *et al.* A method for the solubilization of a (1----3)-beta-D-glucan isolated from *Saccharomyces cerevisiae.* Carbohydr Res 1991; 219: 203-13.

[177]   Diamond MS, Garcia-Aguilar J, Bickford JK, *et al.* The I domain is a major recognition site on the leukocyte integrin Mac-1 (CD11b/CD18) for four distinct adhesion ligands. J Cell Biol 1993; 120: 1031-43.

[178]   Ueda T, Rieu P, Brayer J, *et al.* Identification of the complement iC3b binding site in the beta 2 integrin CR3 (CD11b/CD18). Proc Natl Acad Sci U S A 1994; 91: 10680-4.

[179]   Thornton BP, Vetvicka V, Pitman M, *et al.* Analysis of the sugar specificity and molecular location of the beta-glucan-binding lectin site of complement receptor type 3 (CD11b/CD18). J Immunol 1996; 156: 1235-46.

[180]   Peyron P, Bordier C, N'Diaye EN, Maridonneau-Parini I. Nonopsonic phagocytosis of *Mycobacterium kansasii* by human neutrophils depends on cholesterol and is mediated by CR3 associated with glycosylphosphatidylinositol-anchored proteins. J Immunol 2000; 165: 5186-91.

[181]   Le Cabec V, Cols C, Maridonneau-Parini I. Nonopsonic phagocytosis of zymosan and *Mycobacterium kansasii* by CR3 (CD11b/CD18) involves distinct molecular determinants and is or is not coupled with NADPH oxidase activation. Infect Immun 2000; 68: 4736-45.

[182]   Cywes C, Godenir NL, Hoppe HC, *et al.* Nonopsonic binding of *Mycobacterium tuberculosis* to human complement receptor type 3 expressed in Chinese hamster ovary cells. Infect Immun 1996; 64: 5373-83.

[183]   Ehlers MR. CR3: a general purpose adhesion-recognition receptor essential for innate immunity. Microb Infect / Institut Pasteur 2000; 2: 289-94.

[184]   Tsikitis VL, Albina JE, Reichner JS. Beta-glucan affects leukocyte navigation in a complex chemotactic gradient. Surgery 2004; 136: 384-9.

[185]   Cramer DE, Allendorf DJ, Baran JT, *et al.* Beta-glucan enhances complement-mediated hematopoietic recovery after bone marrow injury. Blood 2006; 107: 835-40.

[186]   Harler MB, Wakshull E, Filardo EJ, *et al.* Promotion of neutrophil chemotaxis through differential regulation of beta 1 and beta 2 integrins. J Immunol 1999; 162: 6792-9.

[187]   Tsikitis VL, Morin NA, Harrington EO, Albina JE, Reichner JS. The lectin-like domain of complement receptor 3 protects endothelial barrier function from activated neutrophils. J Immunol 2004; 173: 1284-91.

[188]   van Bruggen R, Drewniak A, Jansen M, *et al.* Complement receptor 3, not Dectin-1, is the major receptor on human neutrophils for beta-glucan-bearing particles. Mol Immunol 2009; 47: 575-81.

[189]   Li B, Allendorf DJ, Hansen R, *et al.* Yeast beta-glucan amplifies phagocyte killing of iC3b-opsonized tumor cells *via* complement receptor 3-Syk-phosphatidylinositol 3-kinase pathway. J Immunol 2006; 177: 1661-9.

[190]   Vetvicka V, Thornton BP, Ross GD. Soluble beta-glucan polysaccharide binding to the lectin site of neutrophil or natural killer cell complement receptor type 3 (CD11b/CD18) generates a primed state of the receptor capable of mediating cytotoxicity of iC3b-opsonized target cells. J Clin Invest 1996; 98: 50-61.

[191]   Hakomori S. Structure, organization, and function of glycosphingolipids in membrane. Curr Opin Hematol 2003; 10: 16-24.

[192]   Hakomori S, Handa K. Glycosphingolipid-dependent cross-talk between glycosynapses interfacing tumor cells with their host cells: essential basis to define tumor malignancy. FEBS Lett 2002; 531: 88-92.

[193]   Simons K, Toomre D. Lipid rafts and signal transduction. Nat Rev Mol Cell Biol 2000; 1: 31-9.

[194]   Yoshizaki F, Nakayama H, Iwahara C, *et al.* Role of glycosphingolipid-enriched microdomains in innate immunity: microdomain-dependent phagocytic cell functions. Biochim Biophys Acta 2008; 1780: 383-92.

[195]   Zimmerman JW, Lindermuth J, Fish PA, *et al.* A novel carbohydrate-glycosphingolipid interaction between a beta-(1-3)-glucan immunomodulator, PGG-glucan, and lactosylceramide of human leukocytes. J Biol Chem 1998; 273: 22014-20.

[196]   Hahn PY, Evans SE, Kottom TJ, *et al. Pneumocystis carinii* cell wall beta-glucan induces release of macrophage inflammatory protein-2 from alveolar epithelial cells *via* a lactosylceramide-mediated mechanism. J Biol Chem 2003; 278: 2043-50.

[197]   Evans SE, Hahn PY, McCann F, *et al.* Pneumocystis cell wall beta-glucans stimulate alveolar epithelial cell chemokine generation through nuclear factor-kappaB-dependent mechanisms. Am J Resp Cell Mol Biol 2005; 32: 490-7.

[198]   Wakshull E, Brunke-Reese D, Lindermuth J, *et al.* PGG-glucan, a soluble beta-(1,3)-glucan, enhances the oxidative burst response, microbicidal activity, and activates an NF-kappa B-like factor in human PMN: evidence for a glycosphingolipid beta-(1,3)-glucan receptor. Immunopharmacology 1999; 41: 89-107.

[199]   Iwabuchi K, Nagaoka I. Lactosylceramide-enriched glyco-sphingolipid signaling domain mediates superoxide generation from human neutrophils. Blood 2002; 100: 1454-64.

[200]   Nakayama H, Yoshizaki F, Prinetti A, *et al.* Lyn-coupled LacCer-enriched lipid rafts are required for CD11b/CD18-mediated neutrophil phagocytosis of nonopsonized microorganisms. J Leukoc Biol 2008; 83: 728-41.

[201] Sato T, Iwabuchi K, Nagaoka I, *et al.* Induction of human neutrophil chemotaxis by *Candida albicans*-derived beta-1,6-long glycoside side-chain-branched beta-glucan. J Leukoc Biol 2006; 80: 204-11.

[202] Valladeau J, Ravel O, Dezutter-Dambuyant C, *et al.* Langerin, a novel C-type lectin specific to Langerhans cells, is an endocytic receptor that induces the formation of Birbeck granules. Immunity 2000; 12: 71-81.

[203] Stambach NS, Taylor ME. Characterization of carbohydrate recognition by langerin, a C-type lectin of Langerhans cells. Glycobiology 2003; 13: 401-10.

[204] Merad M, Ginhoux F, Collin M. Origin, homeostasis and function of Langerhans cells and other langerin-expressing dendritic cells. Nat Rev Immunol 2008; 8: 935-47.

[205] de Jong MA, Vriend LE, Theelen B, *et al.* C-type lectin Langerin is a beta-glucan receptor on human Langerhans cells that recognizes opportunistic and pathogenic fungi. Mol Immunol 2010; 47: 1216-25.

[206] de Witte L, Nabatov A, Pion M, *et al.* Langerin is a natural barrier to HIV-1 transmission by Langerhans cells. Nature Med 2007; 13: 367-71.

[207] Hunger RE, Sieling PA, Ochoa MT, *et al.* Langerhans cells utilize CD1a and langerin to efficiently present nonpeptide antigens to T cells. J Clin Invest 2004; 113: 701-8.

[208] Takahara K, Omatsu Y, Yashima Y, *et al.* Identification and expression of mouse Langerin (CD207) in dendritic cells. Intl Immunol 2002; 14: 433-44.

[209] Takahara K, Yashima Y, Omatsu Y, *et al.* Functional comparison of the mouse DC-SIGN, SIGNR1, SIGNR3 and Langerin, C-type lectins. Int Immunol 2004; 16: 819-29.

[210] Peiser L, Mukhopadhyay S, Gordon S. Scavenger receptors in innate immunity. Curr Opin Immunol 2002; 14: 123-8.

[211] Rice PJ, Kelley JL, Kogan G, *et al.* Human monocyte scavenger receptors are pattern recognition receptors for (1-->3)-beta-D-glucans. J Leukoc Biol 2002; 72: 140-6.

[212] Dushkin MI, Safina AF, Vereschagin EI, *et al.* Carboxymethylated beta-1,3-glucan inhibits the binding and degradation of acetylated low density lipoproteins in macrophages *in vitro* and modulates their plasma clearance *in vivo*. Cell Biochem Funct 1996; 14: 209-17.

[213] Vereschagin EI, van Lambalgen AA, Dushkin MI, *et al.* Soluble glucan protects against endotoxin shock in the rat: the role of the scavenger receptor. Shock 1998; 9: 193-8.

[214] Pearson A, Lux A, Krieger M. Expression cloning of dSR-CI, a class C macrophage-specific scavenger receptor from Drosophila melanogaster. Proc Natl Acad Sci U S A 1995; 92: 4056-60.

[215] Sen G, Bikah G, Venkataraman C, Bondada S. Negative regulation of antigen receptor-mediated signaling by constitutive association of CD5 with the SHP-1 protein tyrosine phosphatase in B-1 B cells. Eur J Immunol 1999; 29: 3319-28.

[216] Vera J, Fenutria R, Canadas O, *et al.* The CD5 ectodomain interacts with conserved fungal cell wall components and protects from zymosan-induced septic shock-like syndrome. Proc Natl Acad Sci USA 2009; 106: 1506-11.

[217] Ohtani K, Suzuki Y, Eda S, *et al.* The membrane-type collectin CL-P1 is a scavenger receptor on vascular endothelial cells. J Biol Chem 2001; 276: 44222-8.

[218] Nakamura K, Funakoshi H, Tokunaga F, Nakamura T. Molecular cloning of a mouse scavenger receptor with C-type lectin (SRCL)(1), a novel member of the scavenger receptor family. Biochim Biophys Acta 2001; 1522: 53-8.

[219] Nakamura K, Funakoshi H, Miyamoto K, *et al.* Molecular cloning and functional characterization of a human scavenger receptor with C-type lectin (SRCL), a novel member of a scavenger receptor family. Biocheml Biophys Res Comm 2001; 280: 1028-35.

[220] Coombs PJ, Graham SA, Drickamer K, Taylor ME. Selective binding of the scavenger receptor C-type lectin to Lewisx trisaccharide and related glycan ligands. J Biol Chem 2005; 280: 22993-9.

[221] Feinberg H, Taylor ME, Weis WI. Scavenger receptor C-type lectin binds to the leukocyte cell surface glycan Lewis(x) by a novel mechanism. J Biol Chem 2007; 282: 17250-8.

[222] Yoshida T, Tsuruta Y, Iwasaki M, *et al.* SRCL/CL-P1 recognizes GalNAc and a carcinoma-associated antigen, Tn antigen. J Biochem 2003; 133: 271-7.

[223] Nakamura K, Ohya W, Funakoshi H, *et al.* Possible role of scavenger receptor SRCL in the clearance of amyloid-beta in Alzheimer's disease. J Neurosci Res 2006; 84: 874-90.

[224] Jang S, Ohtani K, Fukuoh A, *et al.* Scavenger receptor collectin placenta 1 (CL-P1) predominantly mediates zymosan phagocytosis by human vascular endothelial cells. J Biol Chem 2009; 284: 3956-65.

[225] Febbraio M, Hajjar DP, Silverstein RL. CD36: a class B scavenger receptor involved in angiogenesis, atherosclerosis, inflammation, and lipid metabolism. J Clin Invest 2001; 108: 785-91.

[226] Means TK, Mylonakis E, Tampakakis E, *et al.* Evolutionarily conserved recognition and innate immunity to fungal pathogens by the scavenger receptors SCARF1 and CD36. J Exp Med 2009; 206: 637-53.

[227] Adachi H, Tsujimoto M, Arai H, Inoue K. Expression cloning of a novel scavenger receptor from human endothelial cells. J Biol Chem 1997; 272: 31217-20.

[228]   Tamura Y, Osuga J, Adachi H, *et al.* Scavenger receptor expressed by endothelial cells I (SREC-I) mediates the uptake of acetylated low density lipoproteins by macrophages stimulated with lipopolysaccharide. J Biol Chem 2004; 279: 30938-44.

[229]   Berwin B, Delneste Y, Lovingood RV, *et al.* SREC-I, a type F scavenger receptor, is an endocytic receptor for calreticulin. J Biol Chem 2004; 279: 51250-7.

CHAPTER 4

β-Glucan-Mediated Tumor Immunotherapy – Mechanism of Action and Perspectives

Jun Yan*

Tumor Immunobiology Program, James Graham Brown Cancer, Department of Medicine, University of Louisville, Louisville, KY, USA

Abstract: The beneficial properties of β-glucan in cancer therapy have been recognized for centuries. Their proposed mechanism of action occurs mainly *via* stimulation of macrophages and priming of neutrophil complement receptor 3 (CR3) for eliciting CR3-dependent cellular cytotoxicity of iC3b-opsonized tumor cells. Recent studies revealed that β-glucans may also promote anti-tumor T cell responses. Thus, β-glucan-mediated tumor immunotherapy may engage both the innate and adaptive anti-tumor immune reponses to restrain tumor progression. In this Chapter, we will describe the mechanism of action of β-glucan in cancer immunotherapy and discuss the available data from current β-glucan clinical trials. The potential challenges for this therapy will be also discussed. It is proposed that a biological response modifer β-glucan can serve as a potent adjuvant to specifically modulate both the innate and adaptive immune cells.

INTRODUCTION

Cancer is a leading cause of death worldwide. Approximately 20% of all deaths in the United States are caused by cancer. It has become clear that most cancers have external causes and, in principle, should be preventable. Studies of the association between diet and cancer have demonstrated that the differences in the rates at which various cancers occur in different human populations are often correlated with differences in diet [1,2]. Thus, many efforts have been made to improve diet for preventing cancer. There is accumulating evidence to suggest that the composition of the diet has great impact on our immune system. Therefore, changing dietary compositions as a tool to improve the immune function is a current research focus. β-Glucans, as present in various foods such as cereals and mushrooms, have been widely used as immunostimulating agents to promote immune responses [3]. β-Glucans belong to a group of natural, physiologically active compounds, generally called biological response modifiers (BRMs). Glucans represent highly conserved structural components of cell walls in yeast, fungi, or seaweed. In addition, β-glucans are present in our daily diet as part of the endosperm cell wall in cereals, such as barley and oat. Numerous studies have shown their immuno-stimulating properties, including antitumor activities [3-6].

Despite a long history of research, the exact mechanisms of β-glucan action in cancer therapy remain elusive. In addition, β-glucans from various sources including oat, barley, mushroom, seaweed, some bacteria, and yeast are differential in their structure, conformation, and thus biological activity. This opens another unsolved question- that of which type of β-glucan has superior biological and/or immunological properties? Previous studies demonstrated that soluble, low molecular weight yeast-derived β-glucan bound to a lectin domain within the COOH-terminal region of the CD11b subunit of complement receptor 3 (CR3, CD11b/CD18, $\alpha_m\beta_2$ integrin, Mac-1) [7]. Furthermore, studies have indicated that yeast-derived β-glucans prime neutrophils or natural killer (NK) cells for cytotoxicity against iC3b-opsonized tumors as a result of complement activation by anti-tumor monoclonal antibodies (mAbs) or natural Abs [8, 9]. Dual ligation of neutrophil CR3 mediated by the I-domain ligand, iC3b, and the lectin-like domain (LLD) ligand, β-glucans, leads to degranulation and cytotoxic responses [10]. Thus, β-glucan-mediated tumor immunotherapy utilizes a novel mechanism by which innate immune effector cells are primed to kill iC3b-opsonized tumor cells. Recent studies reveal that β-glucans may also possess regulatory function on adaptive T cell responses. Therefore, β-glucan-mediated tumor immunotherapy may link both innate and adaptive immune responses to fight against cancer.

*****Address correspondence to: Dr. Jun Yan:** Clinical & Translational Research Building, Rm 319, University of Louisville, 505 South Hancock Street, Louisville, KY 40202, Tel: 502 852-3628, Fax: 502 852-2123, Email: jun.yan@louisville.edu

Vaclav Vetvicka and Miroslav Novak (Eds)

β-GLUCAN-MEDIATED TUMOR IMMUNOTHERAPY: INNATE IMMUNITY ARM

Early studies demonstrate that the yeast-derived β-glucan zymosan increases the numbers and function of macrophages and activates the complement system [11,12]. In addition, mushroom β-glucan lentinan was demonstrated to promote lymphokine-activated killer cell activity and NK cell activity [13. 14]. Later studies show that particulate β-glucans induce proinflammatory cytokine secretion thereby stimulating innate immune effector cell activation. These effects appear to be dependent on both toll-like receptor 2 (TLR-2) and dectin-1 pathways [15, 16]. However, soluble yeast-derived β-glucan was demonstrated to bind to the lectin site within the COOH-terminal region of the CD11b subunit of CR3 [17, 19]. CR3 is a β2-integrin family glycoprotein and is widely expressed on the surface of all granulocytes as well as on the surface of monocytes, macrophages, NK cells, and subset of DCs and B cells. It has been demonstrated that CR3-dependent cellular cytotoxicity (CR3-DCC) is one of the critical mechanisms to kill and clear microorganisms [17]. This killing mechanism requires ligation of two distinct binding sites within CR3. Complement activation and iC3b deposition on the microorganisms work in concert with β-glucan, existing in the cell walls of microorganisms, thus leading to CR3-DCC. However, the binding of CR3 to iC3b on tumor cells is not sufficient for leukocytes to kill cancer due to the lack of β-glucan on mammalian cells. Therefore, it was hypothesized that co-administration of complement-activating anti-tumor mAbs with β-glucan would prime CR3 on effector cells for a successful tumoricidal effect. The dual ligation of CR3 by respective ligands on granulocytes, NK cells, and monocytes results in a significant cytotoxicity of tumor cells using an *in vitro* assay system. In addition, significant therapeutic efficacy of combined β-glucan with anti-tumor mAb therapy has been demonstrated in a variety of murine syngeneic tumors [9, 19, 20] as well as in human carcinoma xenograft models [21-25].

To demonstrate critical role of complement activation on tumors and CR3 expression on innate effector cells in β-glucan-mediated tumor immuno-therapy, both C3- and CR3-deficient mice were used to reveal CR3-DCC mechanism *in vivo* for tumor therapy. In an early study, we demonstrated that anti-tumor natural Abs present in mice from a variety of genetic backgrounds were required for tumor regression in response to intravenous (i.v.) administration of a highly purified low MW, soluble zymosan-derived β-glucan poly-saccharide (SZP) [9]. β-Glucan-mediated therapeutic efficacy was significantly compromised in mice with low titers of natural anti-tumor antibody. Similarly, β-glucan therapy was completely failed in Ab-deficient severe combined immunodeficient (SCID) mice. However, β-glucan-mediated therapeutic efficacy was restored by passive immunization with either purified natural Ab in SCID mice or purified anti-tumor mAbs in mice with low natural anti-tumor Ab titers. Similarly, daily i.v. administration of neutral soluble β-glucan (NSG) in addition to anti-tumor mAbs was shown to enhance both tumor regression and long-term survival with respect to treatment with anti-tumor mAb alone or β-glucan alone [19]. However, β-glucan-dependent antitumor responses were significantly abrogated in either C3- or CR3-deficient animals.The rapid clearance of low MW soluble β-glucans limits its clinical utilization [16]. Orally administered β-glucan treatment offers significant advantages such as easy oral administration, relative ease of mass distribution, uncomplicated product storage, long-term product stability, and a relatively low cost. More importantly, orally administered β-glucans, both mushroom derived β-glucan [27, 28] and barley β-glucan [20-22] have demonstrated their therapeutic efficacy in cancer. Pre-clinical animal studies demonstrated orally administered barley β-glucan significantly augmented cytotoxicity of tumors opsonized with anti-tumor mAb and iC3b. This resulted in significant regression and survival benefit in mice bearing either syngeneic tumors or in SCID mice bearing human tumor xenografts [20, 22]. In addition, particulate yeast-derived β-glucan (whole glucan particles, WGPs) were used in pre-clinical animal models in combination with exogenous administration of anti-tumor mAb to test the efficacy for tumor therapy. For example, the combination of daily oral WGPs and weekly administration of an anti-tumor mAb14G2a caused significant tumor regression of 80% or more compared to treatment with mAb alone [20]. In the same experiment, tumor-bearing mice received daily oral barley β-glucan in addition to anti-tumor mAb and the tumor regression induced by the orally administered WGPs was comparable to that of the animals receiving barley β-glucan. However, when observed for long-term survival, 100% of mice receiving oral WGPs in addition to 14G2a lived 8 wks beyond therapy cessation compared to 50% of mice receiving oral barley β-glucan in addition to 14G2a. Further studies are needed to determine whether yeast-derived β-glucans are more efficacious than barley β-glucans. In addition, the tumor regression or survival benefit were significantly diminished in CR3-deficient mice [20]. This important observation confirms similar observations made with i.v. administered low MW soluble β-glucans and reiterates the necessity of the CR3 to bind soluble β-glucan in order to prime CR3 for cytotoxicity. Interestingly, in this particular model, animals receiving WGP alone were observed to have smaller tumors than mice receiving no treatment or mAb alone. This observation is attributed to the persistence of natural Ab directed against either the parent RMA-S cells or to the human MUC1 protein that was expressed on the cells. Alternatively, WGPs may have additional function on activating DCs, thereby eliciting anti-tumor adaptive immunity as we discuss in the next section. Thus, daily oral WGPs were shown to have similar efficacy as soluble β-glucans in terms of both tumor regression and long-term survival, but yielded the convenience of oral dosing.

Immunotherapy with β-glucan substantially enhances the therapeutic efficacy of anti-tumor mAb in the experimental murine breast, lung and lymphoma tumor models. To facilitate translation from preclinical models to clinical application, human carcinoma-challenged xenograft models were further used to test the therapeutic efficacy of combined anti-tumor mAb with β-glucan therapy. In human neuroblastoma xenograft model, daily oral β-glucan treatment in conjunction with anti-GD2 mAb 3F8 produced significant tumor regression and disease stabilization, whereas mAb alone or β-glucan alone did not significantly affect tumor growth [21]. Furthermore, these studies were extended to other human tumor types including melanoma, lymphoma, epidermoid carcinoma, lung carcinoma, ovarian carcinoma, and breast carcinoma [22, 23]. The human non-small cell lung carcinoma (NSCLC) cell line NCI-H23 was implanted in SCID mice to study the therapeutic efficacy of the combined therapy of soluble larger MW PGG β-glucan with humanized anti-epidermal growth factor receptor (EGFR) mAb [24]. In this tumor model, the combined treatment of anti-VEGF mAb with PGG β-glucan successfully inhibited tumor progression and greatly improved long-term tumor-free survival rate. EGFR is widely expressed in the colorectal, head and neck, lung, breast, pancreatic, gastrointestinal carcinomas, and melanoma, is associated with tumor progression, and indicates poor prognosis in patients. Thus, the success of the combinational therapy of PGG β-glucan with anti-EGFR Ab holds great promise for the eradication of tumors and improving the long-term survival for patients whose tumors express high levels of EGFR. In addition, a recent study demonstrated that combinations of anti-EGFR mAbs could trigger complement activation [29], thus potentially promoting β-glucan-mediated therapy. Another recently FDA-approved anti-vascular endothelial growth factor (VEGF) mAb (bevacizumab) was also used to test the combined therapy with PGG β-glucan in human ovarian carcinoma and non-small cell lung carcinoma [25, 30]. Bevacizumab might suppress tumor progression by blocking VEGF-mediated neovascularization, which is necessary for the solid tumor growth. VEGF receptors (VEGFR) are expressed widely in NSCL, leukemia, prostate carcinoma, and breast carcinoma [31-34]. Interestingly, VEGFR have been found on tumor cells as well as in endothelial cells. In addition, a significant fraction of VEGF has also been detected in tumor cells, possibly mediated by its heparin-binding properties or through the binding to VEGFRs. This observation makes bevacizumab a potential candidate that could work in concert with PGG β-glucan immunotherapy to treat cancer patients. This possibility was examined in the SKOV-3 and PC14PE6 carcinoma xenograft models. Both SKOV-3 and PC14PE6 cells express membrane-bound VEGF. Anti-VEGF mAb is a humanized IgG1 Ab that could cause efficient iC3b deposition on tumor cells. Thus, the combination of anti-VEGF with PGG β-glucan indeed induced massive neutrophil infiltration in the tumor and led to enhanced cytotoxicity both *in vitro* and *in vivo* [25, 30]. Thus, co-administration of PGG β-glucan augments the anti-tumor effects of bevacizumab, resulting in synergistic anti-tumor effects. Taken together, these studies have demonstrated the significant therapeutic efficacy using the combination of the complement-activating anti-tumor mAbs with β-glucans from yeast or barley.

The failure of combined β-glucan with anti-tumor mAb therapy in C3- or CR3-deficient mice suggests the importance of complement activation and CR3 for *in vivo* tumoricidal effect [9, 19]. In addition, successful treatment of combined anti-tumor mAbs with β-glucans in SCID or athymic nude mice in human carcinoma xenograft models implies that the main effector cells are innate immune cells. Both *in vivo* and *ex vivo* studies demonstrated that PGG and WGP β-glucans exert their respective effects *via* release of active moiety of β-glucan fragments processed by macrophages [10, 20]. This active moiety β-glucan fragment primes the CR3 of effector cells to elicit CR3-DCC of iC3b-opsonized tumor cells. The signaling event that mediates CR3-DCC has been further characterized [10]. Previous studies demonstrated that protein tyrosine kinase inhibitors genistein or herbimycin A significantly block CR3 priming by β-glucan [35]. Dual ligation of CR3 with active moiety β-glucan fragments and anti-CR3 I-domain mAbs leads to Syk kinase phosphorylation. In addition, the phosphorylated Syk could be co-immunoprecipitated with CD11b, suggesting Syk kinase is phosphorylated and recruited to the CR3 intracellular side [10]. The Syk kinase inhibitor, piceatannol, significantly diminished the dual ligation-mediated increase in PI3K activity, suggesting that PI3 kinase activation is a downstream event after Syk phosphorylation. In addition, both PI3K inhibitor LY 294002 and piceatannol greatly abrogate CR3-DCC, indicating that Syk phosphorylation and the subsequent PI3K activation are the signaling events mediating CR3-DCC. However, it is unknown which kinase phosporylates Syk. A recent study showed that after dual ligation of CD11b CR3 was moved to lipid rafts and interacted with Lyn kinase [36]. LacCer-enriched-lipid rafts were found to be necessary for CR3-mediated phagocytosis of nonopsonized zymosans by human neutrophils. CD11b activation may cause cytoskeletal rearrangement leading to the translocation of CR3 into Lyn-coupled, LacCer-enriched lipid rafts, thus allowing neutrophils to phagocytose nonopsonized zymosans. It remains to be determined whether and how the same mechanism applies to the killing of the iC3b-opsonized tumor cells.

β-GLUCAN-MEDIATED TUMOR IMMUNOTHERAPY: ADAPTIVE IMMUNITY ARM

Although the proposed mechanism of action of β-glucans in cancer therapy occur mainly *via* stimulation of macrophages and priming of neutrophil CR3 for eliciting CR3-dependent cellular cytotoxicity of iC3b-opsonized tumor cells, recent studies demonstrated that β-glucans may also have regulatory functions on adaptive immune responses. β-Glucan-induced T cell differentiation is a very important topic, particularly in the context of a recent study demonstrating that bacterial β-glucan Curdlan is capable of inducing both Th1 and Th17 differentiation [37]. Curdlan also exhibits a potent adjuvant effect on CD8 T cell priming [38]. In addition, Curdlan is capable of converting regulatory T cells (Treg) into Th17 cells *in vitro* [39]. Th17 cells are the third subset of polarized effector T cells characterized by the production of IL-17 and other cytokines and have a critical role in the pathogenesis of several autoimmune diseases [40-44] as well as in anti-tumor T cell responses [45-47]. In mice, the differentiation of Th17 is dependent on the combined action of TGF-β and IL-6 [48, 49]. Recent studies demonstrated that IL-23 and IL-21 are also important for Th17 cell differentiation [50, 51]. Zymosan yeast β-glucans stimulate DCs and macrophages to secrete pro-inflammatory cytokines including IL-12, TNF-α, and IL-6 [16, 27]. However, production of IL-12 and TNF-α was not changed in dectin-1-KO DCs as compared to WT DCs [52]. A recent study shows that zymozan yeast β-glucan also stimulates production of anti-inflammatory cytokines such as IL-10 and TGF-β [53]. *In vivo* treatment of zymosan induces regulatory APCs and Ag-specific T cell tolerance. In addition, β-glucan from *Candida albicans* stimulates human monocyte differentiation into DCs, but these DCs inefficiently polarize naïve T cells [54]. These new emerging data are of great importance and suggest a potential regulatory role of β-glucans in eliciting adaptive T cell responses. In addition, these data postulate that β-glucans from different sources may have differential functions on T cell priming and differentiation. It still needs to be determined whether differential regulatory role of T cell responses is due to conformation differences among different β-glucans or different administration routes of β-glucan therapy.

It is striking that β-glucans could convert Treg into Th17 effector cells [39]. This is significant because the tumor microenvironment contains high frequency of Treg that inhibit effective anti-tumor T cell responses. It is becoming clear that tumors can actively subvert the immune system, and that active immunotherapy is likely defeated as a result of the induction of a variety of immune suppressive mechanisms within the tumor microenvironment [55, 56]. Such suppressive mechanisms are increasingly added, including secretion of immunosuppressive factors (*e.g.* TGF-β and IL-10 secreted by tumor cells or tumor stromal cells) [57], upregulation of negative costimulatory molecules such as CTLA-4, PD-1, and B7-H4 [58, 59], modulation of tryptophan catabolism that induces DC anergy thereby causing T cell tolerance [60], and massive infiltration of Treg and immature or tolerogenic/regulatory DCs within tumors [61]. Along with these factors, myeloid-derived suppressor cells (MDSCs) are also notable as critical contributors to the immune inhibitory signals that limit the adaptive immunity to tumor cells [62 -66]. Therefore, the main challenge of tumor immunotherapy is to modulate the suppressive tumor micro-environment to favor eliciting strong anti-tumor immune responses. Thus, it is critical to determine whether β-glucans have modulatory effect on the tumor microenvironment such as converting Treg into Th17 effector or Th1 cells.

β-GLUCAN CLINICAL TRIALS IN CANCER THERAPY

Since β-glucan therapy has achieved great success in pre-clinical animal models, many efforts have been made to determine their therapeutic efficacy in human patients. Currently, there are several β-glucan clinical trials in cancer therapy (summarized in the Table **1**). Although data from most of these trials have not been released, a recent trial conducted by the Biothera using Imprime PGG plus Erbitux® (cetuximab) and chemodrug has released its clinical results at the 2009 American Society of Clinical Oncology (ASCO) annual meeting [67]. The data suggest that the combination of Imprime PGG®, Erbitux® (cetuximab) and Camptosar® (irinotecan) nearly doubled the overall response rate of second- and third-line metastatic colorectal cancer patients compared with cetuximab and irinotecan treatment. Median progression-free survival increased 38% to 22 weeks with the Imprime PGG combination, compared with 16 weeks for the standard of care. The overall response rate (complete response + partial response) was 30% with the Imprime PGG combination, compared to 16% for the standard of care. The disease control rate (complete response + partial response + stable disease) was 100% for the Imprime PGG group, compared with 61% for the standard of care. These initial clinical results are very promising and exciting. However, caution is needed due to low number of patients enrolled in this trial. Regardless, these data warrant further clinical investigations to truly test its efficacy in cancer therapy.

Table I: β-Glucan clinical trials in cancer therapy

Tumor Type	Intervention	Phase	Sponsor	Status/citation
Neuroblastoma	β-glucan plus anti-GD2 mAb 3F8	Phase I	Memorial Sloan-Kettering Cancer Center	Completed http://clinicaltrials.gov/ct2/show/NCT00037011
Neuroblastoma	Oral β-glucan plus a vaccine containing two antigens (GD2L and GD3L) covalently linked to KLH	Phase I	Memorial Sloan-Kettering Cancer Center	Recruiting http://clinicaltrials.gov/ct2/show/NCT00911560?term=glucan+trials&rank=1
Chronic Lymphocytic Leukemia(CLL)/Small Lymphocytic Lymphoma (SLL)	Rutiximab plus oral glucan supplement	Phase II	James Graham Brown Cancer Center	Completed http://clinicaltrials.gov/ct2/show/NCT00290407?term=glucan+trials&rank=8
Advanced non-small cell lung carcinoma	β-Glucan MM10-001	Phase I	Beckman Research Institute	Recruiting http://clinicaltrials.gov/ct2/show/NCT00857025?term=glucan+trials&rank=13
Relapsed or Progressive Lymphoma or Leukemia, or Lymphoproliferative Disorder Related to Donor Stem Cell Transplantation	Rutiximab plus oral β-glucan	Phase I	Memorial Sloan-Kettering Cancer Center	Completed http://clinicaltrials.gov/ct2/show/NCT00087009?term=glucan+trials&rank=15
Breast cancer	Soluble β-glucan plus mAb plus chemotherapy	Phase I/II	Biotec Pharmacon ASA	Completed http://clinicaltrials.gov/ct2/show/NCT00533364?term=glucan+trials&rank=17
Non-Hodgkin's Lymphoma	Soluble β-glucan plus Rutuximab plus COP/CHOP	Phase I	Biotec Pharmacon ASA	Completed http://clinicaltrials.gov/ct2/show/NCT00533728?term=glucan+trials&rank=20
Non-small cell lung carcinoma	Oral β-glucan supplement	N/A	James Graham Brown Cancer Center	Recruiting http://clinicaltrials.gov/ct2/show/NCT00682032?term=glucan+trials&rank=25
Non-small cell lung carcinoma	Imprime PGG in combination with Avastin® (bevacizumab) and two chemotherapeutic agents, carboplatin and paclitaxel.	Phase II	Biothera	Recruiting http://www.biotherapharma.com/pharmaceutical/BiotheraLungCancerTrials.htm
Non-small cell lung carcinoma	Imprime PGG in combination with Erbitux® (cetuximab), carboplatin and paclitaxel	Phase II	Biothera	Recruiting http://www.biotherapharma.com/pharmaceutical/BiotheraLungCancerTrials.htm
Metastatic colorectal cancer	Imprime PGG™ and Erbitux®	Phase Ib/IIa	Biothera	Recruiting http://www.biotherapharma.com/pharmaceutical/BiotherainitiatesPhaseIbIIaclinicaltrialincancerpatients.html
KRAS-Mutated Colorectal Cancer Patients	Imprime PGG™ and Erbitux®	Phase II	Biothera	Recruiting http://www.biotherapharma.com/pharmaceutical/KRAS.htm

PERSPECTIVES

Currently, most anti-tumor immunotherapy strategies except anti-tumor mAb therapy are reserved for advanced patients that have failed conventional therapy. The challenges of antitumor immunotherapy lie in many aspects including immune tolerance established by tumors. Although robust populations of immune effector cells can be generated *ex vivo*, clinical and pathologic complete responses remain rare in cancer patients treated with these modalities. Combined immunotherapy utilizing β-glucan and anti-tumor mAbs is one means of breaking immune tolerance to tumors. This combination therapy offers several unique advantages over other immunotherapeutic approaches. For example, β-glucan therapy can use humanized mAbs to target tumors and thus does not rely on the patient's own immune responses, which are frequently suppressed. Any anti-tumor mAb that is capable of activating complement could be used in combination with β-glucan. The clinical usage of anti-tumor mAbs continues to grow and will be incorporated into the standard of care for cancers of multiple organs. Increases in the number of anti-tumor Abs offer more opportunities to design versatile combinations with β-glucan therapy. This modality can be also used in synergy with most tumor vaccines as long as the antitumor humoral responses are elicited and the Abs can bind to tumors resulting in complement activation. Even with tumor vaccines that generate non-protective Ab responses and fail in clinical trial [68-70], it is still feasible for the vaccines to be used in conjunction with β-glucan if these Abs are capable of binding to tumors and activating complement. In addition, β-glucans can serve as a potent adjuvant to regulate anti-tumor T cell responses. For those tumor vaccines that elicit potent cytotoxic T lymphocyte (CTL) responses as well as humoral responses [71], combined therapy with β-glucan could add another protective layer of innate anti-tumor immunity to the adaptive anti-tumor immunity these tumor vaccines generate. If the effect of converting regulatory T cells into Th17 effector cells by β-glucans is confirmed in *in vivo* tumor models, glucan therapy may possess an ability of modulating the suppressive tumor microenvironment to favor anti-tumor immunotherapy. Therefore, β-glucan-mediated therapy could not only target innate granulocytes to elicit CR3-DCC but also stimulate anti-tumor T cell responses to provide a more effective means of eliminating tumors and developing a long-term tumor-specific T cell immunity that prevents tumor recurrence.

Intravenous administration of soluble β-glucan PGG has demonstrated a strong safety profile. PGG β-glucan does not induce any proinflammatory cytokines *in vitro*. In addition, only 10% of the marginated pool of neutrophils is primed for CR3-DCC *in vivo* [72]. This may represent an internal threshold to limit the unchecked activation of neutrophils that may lead to autoimmunity or unnecessary inflammation. Given the high percentage and high renewal rate of neutrophils, this priming may result in a robust, persistent effector population capable of mediating *in vivo* tumor regression and survival.

Despite the demonstration of β-glucan-primed neutrophils that mediate CR3-DCC of iC3b-opsonized tumors *ex vivo*, there are some challenges to the success of combined β-glucan with anti-tumor mAb therapy. For example, the anti-inflammatory milieu of the tumor microenvironment and the over-expression of membrane complement regulatory proteins (mCPRs) limit the capacity of β-glucan-primed neutrophil to mediate tumor regression *in vivo*. Anti-inflammatory cytokines secreted by the tumor and tumor stroma may create a barrier to block the entry of β-glucan-primed neutrophils. Over-expression of CD55 on SKOV-3 tumors significantly impaired complement activation and C5a release within the tumors, resulting in the paucity of tumor-infiltrating neutrophils [24]. Inefficiency of iC3b deposition on tumors and C5a release within the tumors may cause therapeutic failure of combination therapy. Some potential means to overcome these obstacles include breaking the physiological barrier to inflammation by utilizing exogenously administered pro-inflammatory cytokines or inducing the tumor cells or stroma to produce inflammatory cytokines although caution is needed since chronic inflammation promotes tumor progression [73]. Similarly, strategies that result in the deposition of more iC3b, including anti-tumor cocktail mAbs that bind to multi-targets of TAA [29, 74], amplification of complement activation, or manipulation of mCRPs would be expected to improve β-glucan-mediated therapeutic efficacy. These studies should shed the light on how to better design robust and effective combination therapy in cancer.

REFERENCES

[1]   Bode AM, Dong Z. Cancer prevention research - then and now. Nat Rev Cancer 2009; 9:508-16.
[2]   Rajamanickam S, Agarwal R. Natural products and colon cancer: current status and future prospects. Drug Dev Res 2008; 69:460-71.

[3]   Wasser SP, Weis AL. Therapeutic effects of substances occurring in higher Basidiomycetes mushrooms: a modern perspective. Crit Rev Immunol 1999; 19:65-96.

[4]   Hunter JT, Meltzer MS, Ribi E, *et al.* Glucan: attempts to demonstrate therapeutic activity against five syngeneic tumors in guinea pigs and mice. J Natl Cancer Inst 1978; 60:419-24.

[5]   Vetvicka V, Vashishta A, Saraswat-Ohri S, Vetvickova J. Immunological effects of yeast- and mushroom-derived beta-glucans J Med Food. 2008; 11:615-22.

[6]   Novak M, Vetvicka V. Glucans as biological response modifiers. Endocr Metab Immune Disord Drug Targets 2009; 9:67-75.

[7]   Xia Y, Vetvicka V, Yan J, Hanikyrova M, Mayadas T, Ross GD. The beta-glucan-binding lectin site of mouse CR3 (CD11b/CD18) and its function in generating a primed state of the receptor that mediates cytotoxic activation in response to iC3b-opsonized target cells. J Immunol 1999; 162:2281-90.

[8]   Vetvicka V, Thornton BP, Wieman TJ, Ross GD. Targeting of natural killer cells to mammary carcinoma *via* naturally occurring tumor cell-bound iC3b and beta-glucan-primed CR3 (CD11b/CD18). J Immunol 1997; 159:599-605.

[9]   Yan J, Vetvicka V, Xia Y, *et al.* Beta-glucan, a "specific" biologic response modifier that uses antibodies to target tumors for cytotoxic recognition by leukocyte complement receptor type 3 (CD11b/CD18). J Immunol 1999; 163:3045-52.

[10]   Li B, Allendorf DJ, Hansen R, *et al.* Yeast beta-glucan amplifies phagocyte killing of iC3b-opsonized tumor cells *via* complement receptor 3-Syk-phosphatidylinositol 3-kinase pathway J Immunol 2006; 177:1661-9.

[11]   Kokoshis PL, Williams DL, Cook JA, Di Luzio NR. Increased resistance to Staphylococcus aureus infection and enhancement in serum lysozyme activity by glucan. Science 1978; 199:1340-2.

[12]   Ezekowitz RA, Sim RB, Hill M, Gordon S. Local opsonization by secreted macrophage complement components. Role of receptors for complement in uptake of zymosan. J Exp Med 1984; 159:244-60.

[13]   Tani M, Tanimura H, Yamaue H, *et al.* Augmentation of lymphokine-activated killer cell activity by lentinan. Anticancer Res 1993; 13:1773-6.

[14]   Tani M, Tanimura H, Yamaue H, *et al. In vitro* generation of activated natural killer cells and cytotoxic macrophages with lentinan. Eur J Clin Pharmacol 1992; 42:623-7.

[15]   Brown GD, Herre J, Williams DL, Willment JA, Marshall AS, Gordon S. Dectin-1 mediates the biological effects of beta-glucans. J Exp Med 2003; 197:1119-24.

[16]   Gantner BN, Simmons RM, Canavera SJ, Akira S, Underhill DM. Collaborative induction of inflammatory responses by dectin-1 and Toll-like receptor 2. J Exp Med 2003; 197:1107-17.

[17]   Ross GD, Cain JA, Myones BL, Newman SL, Lachmann PJ. Specificity of membrane complement receptor type three (CR3) for beta-glucans. Complement 1987; 4:61-74.

[18]   Thornton BP, Vetvicka V, Pitman M, Goldman RC, Ross GD. Analysis of the sugar specificity and molecular location of the beta-glucan-binding lectin site of complement receptor type 3 (CD11b/CD18). J Immunol 1996; 156:1235-46.

[19]   Hong F, Hansen RD, Yan J, *et al.* Beta-glucan functions as an adjuvant for monoclonal antibody immunotherapy by recruiting tumoricidal granulocytes as killer cells. Cancer Res 2003; 63:9023-31.

[20]   Hong F, Yan J, Baran JT, Allendorf DJ, *et al.* Mechanism by which orally administered beta-1,3-glucans enhance the tumoricidal activity of antitumor monoclonal antibodies in murine tumor models. J Immunol 2004; 173:797-806.

[21]   Cheung NK, Modak S. Oral (1-->3),(1-->4)-beta-D-glucan synergizes with antiganglioside GD2 monoclonal antibody 3F8 in the therapy of neuroblastoma. Clin Cancer Res 2002; 8:1217-23.

[22]   Cheung NK, Modak S, Vickers A, Knuckles B. Orally administered beta-glucans enhance anti-tumor effects of monoclonal antibodies. Cancer Immunol Immunother 2002; 51:557-64.

[23]   Modak S, Koehne G, Vickers A, O'Reilly RJ, Cheung NK. Rituximab therapy of lymphoma is enhanced by orally administered (1-->3),(1-->4)-D-beta-glucan. Leuk Res 2005; 29:679-83.

[24]   Li B, Allendorf DJ, Hansen R, *et al.* Combined yeast β-glucan and antitumor monoclonal antibody therapy requires C5a-mediated neutrophil chemotaxis *via* regulation of decay-accelerating factor CD55. Cancer Res 2007; 67:7421-30.

[25]   Salvador C, Li B, Hansen R, Cramer DE, Kong M, Yan J. Yeast-Derived β-glucan augments the therapeutic efficacy mediated by anti-vascular endothelial growth factor monoclonal antibody in human carcinoma xenograft models. Clin Cancer Res 2008; 14:1239-47.

[26]   Yan J, Vetvicka V, Xia Y, Hanikyrova M, Mayadas TN, Ross GD. Critical role of Kupffer cell CR3 (CD11b/CD18) in the clearance of IgM-opsonized erythrocytes or soluble beta-glucan. Immunopharmacology 2000; 46:39-54.

[27]   Nanba H, Kuroda H. Antitumor mechanisms of orally administered shiitake fruit bodies. Chem Pharm Bull (Tokyo) 1987; 35:2459-64.

[28]   Suzuki I, Sakurai T, Hashimoto K, *et al.* Inhibition of experimental pulmonary metastasis of Lewis lung carcinoma by orally administered beta-glucan in mice. Chem Pharm Bull (Tokyo) 1991; 39:1606-8.

[29]   Dechant M, Weisner W, Berger S, *et al.* Complement-dependent tumor cell lysis triggered by combinations of epidermal growth factor receptor antibodies. Cancer Res 2008; 68:4998-5003.

[30]   Zhong W, Hansen R, Li B, *et al.* Effect of yeast-derived beta-glucan in conjunction with bevacizumab for the treatment of human lung adenocarcinoma in subcutaneous and orthotopic xenograft models. J Immunother 2009; 32:703-12.

[31]   Decaussin M, Sartelet H, Robert C, *et al.* Expression of vascular endothelial growth factor (VEGF) and its two receptors (VEGF-R1-Flt1 and VEGF-R2-Flk1/KDR) in non-small cell lung carcinomas (NSCLCs): correlation with angiogenesis and survival. J Pathol 1999; 188:369-77.

[32]   Bellamy WT, Richter L, Frutiger Y, Grogan TM. Expression of vascular endothelial growth factor and its receptors in hematopoietic malignancies. Cancer Res 1999; 59:728-33.

[33]   Ferrer FA, Miller LJ, Lindquist R, Kowalczyk P, Laudone VP, Albertsen PC, *et al.* Expression of vascular endothelial growth factor receptors in human prostate cancer. Urology 1999; 54:567-72.

[34]   Price DJ, Miralem T, Jiang S, Steinberg R, Avraham H. Role of vascular endothelial growth factor in the stimulation of cellular invasion and signaling of breast cancer cells. Cell Growth Differ 2001; 12:129-35.

[35]   Vetvicka V, Thornton BP, Ross GD. Soluble beta-glucan polysaccharide binding to the lectin site of neutrophil or natural killer cell complement receptor type 3 (CD11b/CD18) generates a primed state of the receptor capable of mediating cytotoxicity of iC3b-opsonized target cells. J Clin Invest 1996; 98:50-61.

[36]   Nakayama H, Yoshizaki F, Prinetti A, *et al.* Lyn-coupled LacCer-enriched lipid rafts are required for CD11b/CD18-mediated neutrophil phagocytosis of nonopsonized microorganisms. J Leukoc Biol 2008; 83:728-41.

[37]   LeibundGut-Landmann S, Gross O, Robinson MJ, *et al.* Syk- and CARD9-dependent coupling of innate immunity to the induction of T helper cells that produce interleukin 17. Nat Immunol 2007; 8:630-8.

[38]   Leibundgut-Landmann S, Osorio F, Brown GD, Reis e Sousa C. Stimulation of dendritic cells *via* the dectin-1/Syk pathway allows priming of cytotoxic T-cell responses. Blood 2008; 112:4971-80.

[39]   Osorio F, LeibundGut-Landmann S, Lochner M, *et al.* DC activated *via* dectin-1 convert Treg into IL-17 producers. Eur J Immunol 2008; 38:3274-81.

[40]   Dong C. Diversification of T-helper-cell lineages: finding the family root of IL-17-producing cells. Nat Rev Immunol 2006; 6:329-33.

[41]   Weaver CT, Hatton RD, Mangan PR, Harrington LE. IL-17 family cytokines and the expanding diversity of effector T cell lineages. Annu Rev Immunol 2007; 25:821-52.

[42]   Weaver CT, Harrington LE, Mangan PR, Gavrieli M, Murphy KM. Th17: an effector CD4 T cell lineage with regulatory T cell ties. Immunity 2006; 24:677-88.

[43]   Park H, Li Z, Yang XO, *et al.* A distinct lineage of CD4 T cells regulates tissue inflammation by producing interleukin 17. Nat Immunol 2005; 6:1133-41.

[44]   Harrington LE, Hatton RD, Mangan PR, *et al.* Interleukin 17-producing CD4+ effector T cells develop *via* a lineage distinct from the T helper type 1 and 2 lineages. Nat Immunol 2005; 6:1123-32.

[45]   Martin-Orozco N, Muranski P, Chung Y, *et al.* T helper 17 cells promote cytotoxic T cell activation in tumor immunity. Immunity 2009; 31:787-98.

[46]   Kryczek I, Banerjee M, Cheng P, *et al.* Phenotype, distribution, generation, and functional and clinical relevance of Th17 cells in the human tumor environments. Blood 2009; 114:1141-9.

[47]   Kryczek I, Wei S, Szeliga W, Vatan L, Zou W. Endogenous IL-17 contributes to reduced tumor growth and metastasis. Blood 2009;114:357-9.

[48]   Mangan PR, Harrington LE, O'Quinn DB, *et al.* Transforming growth factor-beta induces development of the T(H)17 lineage. Nature 2006; 441:231-4.

[49]   Bettelli E, Carrier Y, Gao W, *et al.* Reciprocal developmental pathways for the generation of pathogenic effector TH17 and regulatory T cells. Nature 2006; 441:235-8.

[50]   Korn T, Bettelli E, Gao W, *et al.* IL-21 initiates an alternative pathway to induce proinflammatory T(H)17 cells. Nature 2007; 448:484-7.

[51]   Nurieva R, Yang XO, Martinez G, *et al.* Essential autocrine regulation by IL-21 in the generation of inflammatory T cells. Nature 2007; 448:480-3.

[52]   Taylor PR, Tsoni SV, Willment JA, *et al.* Dectin-1 is required for beta-glucan recognition and control of fungal infection. Nat Immunol 2007; 8:31-8.

[53]   Dillon S, Agrawal S, Banerjee K, *et al.* Yeast zymosan, a stimulus for TLR2 and dectin-1, induces regulatory antigen-presenting cells and immunological tolerance. J Clin Invest 2006; 116:916-28.

[54]   Nisini R, Torosantucci A, Romagnoli G, *et al.* β-Glucan of Candida albicans cell wall causes the subversion of human monocyte differentiation into dendritic cells. J Leukoc Biol 2007; 82:1136-42.

[55] Croci DO, Zacarias Fluck MF, Rico MJ, Matar P, Rabinovich GA, Scharovsky OG. Dynamic cross-talk between tumor and immune cells in orchestrating the immunosuppressive network at the tumor microenvironment. Cancer Immunol Immunother 2007; 56:1687-700.

[56] Rabinovich GA, Gabrilovich D, Sotomayor EM. Immunosuppressive strategies that are mediated by tumor cells. Annu Rev Immunol 2007; 25:267-96.

[57] Wrzesinski SH, Wan YY, Flavell RA. Transforming Growth Factor-β and the Immune Response: Implications for Anticancer Therapy. Clin Cancer Res 2007; 13:5262-70.

[58] Zang X, Allison JP. The b7 family and cancer therapy: costimulation and coinhibition. Clin Cancer Res 2007; 13:5271-9.

[59] Kryczek I, Wei S, Zhu G, *et al.* Relationship between B7-H4, regulatory T cells, and patient outcome in human ovarian carcinoma. Cancer Res 2007; 67:8900-5.

[60] Gajewski TF. Failure at the effector phase: immune barriers at the level of the melanoma tumor microenvironment. Clin Cancer Res 2007; 13:5256-61.

[61] Zou W. Regulatory T cells, tumour immunity and immunotherapy. Nat Rev Immunol 2006; 6:295-307.

[62] Gabrilovich DI, Bronte V, Chen SH, *et al.* The terminology issue for myeloid-derived suppressor cells. Cancer Res 2007;67:425;

[63] Bunt SK, Yang L, Sinha P, Clements VK, Leips J, Ostrand-Rosenberg S. Reduced inflammation in the tumor microenvironment delays the accumulation of myeloid-derived suppressor cells and limits tumor progression. Cancer Res 2007; 67:10019-26.

[64] Sinha P, Clements VK, Bunt SK, Albelda SM, Ostrand-Rosenberg S. Cross-talk between myeloid-derived suppressor cells and macrophages subverts tumor immunity toward a type 2 response. J Immunol 2007; 179:977-83.

[65] Huang B, Pan PY, Li Q, *et al.* Gr-1+CD115+ immature myeloid suppressor cells mediate the development of tumor-induced T regulatory cells and T-cell anergy in tumor-bearing host. Cancer Res 2006; 66:1123-31.

[66] Pan PY, Wang GX, Yin B, *et al.* Reversion of immune tolerance in advanced malignancy: modulation of myeloid derived suppressor cell development by blockade of SCF function. Blood 2007; 111:219-28.

[67] M. E. Tamayo GHC, J. B. Bautista, M. L. Flores, M. R. Kurman, M. M. Paul, M. A. Gargano, M. L. Patchen. A phase Ib/2, dose-escalating, safety, and efficacy study of imprime PGG, cetuximab and irinotecan in patients with advanced colorectal cancer (CRC). J Clin Oncol 2009; 27: e15062.

[68] Foon KA, Chakraborty M, John WJ, Sherratt A, Kohler H, Bhattacharya-Chatterjee M. Immune response to the carcinoembryonic antigen in patients treated with an anti-idiotype antibody vaccine. J Clin Invest 1995; 96:334-42.

[69] Karanikas V, Hwang LA, Pearson J, *et al.* Antibody and T cell responses of patients with adenocarcinoma immunized with mannan-MUC1 fusion protein. J Clin Invest 1997; 100:2783-92.

[70] Nicholson S, Bomphray CC, Thomas H, *et al.* A phase I trial of idiotypic vaccination with HMFG1 in ovarian cancer. Cancer Immunol Immunother 2004; 53:809-16.

[71] Boscardin SB, Hafalla JC, Masilamani RF, *et al.,* Antigen targeting to dendritic cells elicits long-lived T cell help for antibody responses. J Exp Med 2006; 203:599-606.

[72] Allendorf DJ, Yan J, Ross GD, *et al.* C5a-Mediated leukotriene B4-amplified neutrophil chemotaxis is essential in tumor immunotherapy facilitated by anti-tumor monoclonal antibody and β-glucan. J Immunol. 2005; 174:7050-6.

[73] Coussens LM, Werb Z. Inflammation and cancer. Nature 2002; 420:860-7.

[74] Spiridon CI, Ghetie MA, Uhr J, *et al.* Targeting multiple Her-2 epitopes with monoclonal antibodies results in improved antigrowth activity of a human breast cancer cell line *in vitro* and *in vivo*. Clin Cancer Res 2002; 8:1720-30.

Biology and Chemistry of Beta Glucan, Vol. 01, 2011, 48-67

CHAPTER 5

Use of β-Glucans for Drug Delivery Applications

Ernesto Soto and Gary Ostroff*

Program in Molecular Medicine, University of Massachusetts Medical School, Worcester, MA, USA

Abstract: β(1-3)-D-Glucans are a structurally varied class of polysaccharides that have been long studied for their immunomodulatory and physical properties. In this review we focus on the work intersecting these two properties, the use of β(1-3)-D-glucans for drug delivery. Natural and chemically-modified β(1-3)-D-glucans have been used for drug delivery in many forms: 1) release of drugs from a glucan gel matrix, 2) in combination with other materials to form suitable drug delivery systems, 3) as carriers in drug formulations consisting of gels, tablets and ingestible films, 4) as soluble and particulate conjugates, and 5) and as soluble and particulate β(1-3)-D-glucans for encapsulation and delivery of macromolecules. The scope of this review is to summarize the significant developments using soluble and insoluble β(1-3)-D-glucans for drug delivery reported in the literature.

INTRODUCTION

β(1-3)-D-Glucans are a class of polysaccharides consisting of D-glucose monomers naturally found in fungi, algae, plants and some bacteria, and are the major pathogen associated molecular pattern (PAMPs) of fungi, broadly recognized phylogenetically. β(1-3)-D-Glucans have been studied extensively as an immune-stimulant in anti-infective, anti-tumor, immunoadjuvant in cancer therapy, wound healing, and for cholesterol lowering and many other therapeutic applications [1, 2].

The diversity of β(1-3)-D-glucan structures comes from the different natural sources, degree of branching and glycosidic bonds, wide range of molecular weights, and three-dimensional configuration (single strand chains, double or triple stranded helix chains). All these variables have a significant effect on both physical properties, such as solubility and rheology, and biological activity. The major classes of β(1-3)-D-glucan are illustrated in Table **1**.

Table 1: Examples of β-glucans

Name	Source	Chemical structure showing glycosidic bond
Curdlan [2]	Bacteria	Linear 1-3
Oat glucan [32]	Cereal	1-3 1-4 coblock linear

*Address correspondence to: Dr. Gary Ostroff: Program in Molecular Medicine, University of Massachusetts Medical School, 373 Plantation Street, Worcester, MA 01604; E-mail: gostroff@umassmed.edu

Vaclav Vetvicka and Miroslav Novak (Eds)

Name	Source	Chemical structure showing glycosidic bond
Scleroglucan [2]	Mushroom	 1-6 branched 1-3 linear
PGG-Glucan [33]	Yeast	 1-6 branched 1-3 linear

BIOLOGICAL ACTIVITIES OF β(1-3)-D-GLUCANS

The most studied application for β(1-3)-D-glucans is based on their ability to activate the immune system. Two receptors, dectin-1 and the Complement Receptor 3 (CR3 or CD11b/CD18) present on the surface of innate immune cells have been found to be responsible for binding to β(1-3)-D-glucans [3] and the signaling mediated by both receptors have been characterized at the molecular level [4-6]. Dectin-1 is primarily expressed on macrophages and dendritic cells, and is largely absent on NK cells. CR3 is expressed largely on neutrophils, monocytes, and natural killer (NK) cells, and is less abundant on macrophages [7].

A significant difference between β(1-3)-D-glucans and other PAMPs is that glucans do not over stimulate the immune system as is seen with other immunomodulators, such as the bacterial cell wall products, lipopolysaccharide, peptidoglycan, and lipopeptides, which are contraindicated in individuals with autoimmune diseases or allergies [8, 9]. Although many of the original studies with β(1-3)-D-glucans were done *via* parenteral administration, more recent studies have demonstrated the possibility for oral administration without compromising biological activity [10-12].

β(1-3)-D-Glucans have also been used for their immunomodulatory effect in topical formulations for wound healing and adjuvants in varied pharmaceutical products, and also in studies for prevention of infections and treatment of a wide variety of different diseases [13]. Glucans present in natural products such as barley and oat cereals, and edible mushrooms have been shown to decrease levels of serum cholesterol and liver low-density lipoproteins leading to reduction in atherosclerosis and cardiovascular disease hazards [14]. Some of the most significant applications of β(1-3)-D-glucans are summarized below. More detailed information can be found in recent reviews [1, 15] and in chapter 1 in this book.

Prevention of Infection

Betafectin or PGG-glucan, a soluble β(1-6) branched β(1-3) glucan purified from the cell walls of *S. cerevisiae,* is a macrophage-specific anti-infective immunomodulator that was evaluated in human clinical trials in the 1990's. Results of phase II trials showed that injectable PGG-glucan reduced post-surgery infections by 39% [16-18].

Unfortunately, these results were not confirmed in large-scale Phase III clinical trials [16]. Oral consumption of whole glucan particles (WGP) also provided a strong enhancement of natural immunity. For example oral administration of WGP given with or without antibiotics protected mice against anthrax infection, while a control group of mice treated only with antibiotic did not survive [19].

Use in Cancer Treatments

In Japan, β(1-3)-D-glucans like Lentinan derived from the Shiitake mushroom, and Polysaccharide K derived from *Coriolus versicolor* are approved for use as immunoadjuvants for cancer therapy and have been used for over 20 years [20-23]. In human patients with advanced gastric and colorectal cancer administration of glucans derived from shiitake mushrooms and chemotherapy resulted in prolonged survival times [24-26].

Several studies have demonstrated that oral forms of yeast β(1-3)-D-glucan have a similar protective effect as injected forms, including defense against infectious disease and cancer [19, 27-30]. β(1-3)-glucan in conjuction with interferon gamma has also been shown to inhibit tumors and liver metastesis in mice [31]. Recently, orally-delivered glucan was found to significantly increase proliferation and activation of monocytes in peripheral blood of patients with advanced breast cancer [34]. Low molecular weight glucans from *Aureobasidium pullulans* (black yeast) also show antitumor and antimetastatic activity [35].

Preclinical studies have shown that a soluble yeast β(1-3)-D-glucan product, Imprime PGG, when used in combination with certain monoclonal antibodies or cancer vaccines, offers significant improvements in long-term survival versus monoclonal antibodies alone [3].

Wound Healing

Therapy with β(1-3)-D-glucan has provided improvements, such as fewer infections, reduced mortality and stronger tensile strength of scar tissue, and activation of tissue regeneration process. β(1-3)-D-Glucans have been used for diverse applications in wound healing, for example topical compositions for treatments of burns and scarring [36], and cosmetic applications due to its immunomodulatory properties.

Radiation Exposure

Several studies have demonstrated the effect of β(1-3)-D-glucan to enhance recovery after radiation exposure. Hematopoietic activity was first demonstrated with β(1-3)-D-glucan in the 1980's [37]. Mice exposed to gamma radiation and treated intravenously (IV) with particulate or soluble β(1-3)-D-glucan exhibited a significant enhanced recovery of blood leukocyte, platelet and red blood cell counts [38]. Other studies have also shown the effect of β(1-3)-D-glucan to reverse adverse effects from chemotherapeutic drugs such as fluorouracil [16], carboplatin and cyclophosphamide [39].

Treatment of Allergies

Orally administered yeast-β(1-3)-D-glucan has been used in treatment of allergic rhinitis, a disease caused by an IgE-mediated allergic inflammation of the nasal mucosa. Glucan administration decreased the levels of IL-4 and IL-5 cytokines responsible for the clinical manifestation of this disease, while increased the levels of IL-12. Based on these studies, glucan may have a role as an adjuvant to standard treatment in patients with allergic diseases [40].

Intestinal Infections

β(1-3)-D-Glucan has been found to protect against septic infections produced by bacteria such as *E. coli* or *S. aureus* [41, 42]. This is likely the result of yeast β(1-3)-D-glucan's ability to prime leukocytes to more easily locate and kill non-self cells including bacteria. Another proposed mechanism is that glucan protects against oxidative organ injury [43]. Oral delivery and gastrointestinal absorption of glucans stimulate resistance to infectious challenge with *Staphylococcus aureus* and *Candida albicans* [28]. Cereal, mushroom and yeast glucans facilitate bowel motility and can be used in amelioration of intestinal problems [44, 45]. Non-digestible β(1-3)-D-glucan have been reported to modulate mucosal immunity of the intestinal tract [46].

USE OF β-GLUCANS FOR DRUG DELIVERY

Glucans have been used in drug delivery systems either as an actual drug carrier, an adjuvant, or in combination with other materials to form suitable drug delivery systems (*i.e.,* nanogels [47] and as stabilizer in microcapsules and nanocapsules [48]). This chapter presents the most significant applications of β(1-3)-D-glucan for drug delivery in four sections: (1) use of β(1-3)-D-glucan (primarily curdlan, scleroglucan and their synthetic derivatives) as carriers in drug formulations consisting of gels, tablets and ingestible films; (2) preparation of particles coated with β(1-3)-D-glucan; (3) the use of soluble β(1-3)-D-glucan for encapsulation and delivery of macromolecules (*i.e.,* DNA delivery with schizophyllan complexes); and (4) the development of the yeast β(1-3)-D-glucan particle drug delivery technology.

Use of β(1-3)-D-Glucan as Carriers in Gels and other Drug Formulations

β(1-3)-D-glucans have been used in drug formulations for both their immunoadjuvant properties and as an excipient to stabilize drug formulations, facilitate drug delivery and controlled drug release. For example, pullulan was used in combination with water soluble polymers (modified cellulose, starch, and carrageenan) to prepare ingestible films that can contain pharmaceutical, cosmetic, or biologically active agents [49]. Other materials, such as oat glucan have been reported as a carrier for chemical substances through the skin [50]. Conjugates of moxifloxacin and carboxymethylated glucan have been shown to have enhanced activity against intracellular *Mycobacterium tuberculosis* [51].

Drug Delivery Applications with Curdlan

Curdlan is a high molecular weight glucan composed entirely of β(1-3)-D-glucosidic linkages that exists as a triple helical in solution. It is produced as an exopolysaccharide by the bacteria *Agrobacterium biobar* and *Alcaligenes faecalis*. Curdlan has the property of forming elastic gels upon heating in slightly alkaline (0.01 M NaOH) aqueous suspensions.

Curdlan has been investigated for preparation of tablets containing theophylline, a drug used in treatment of respiratory diseases such as COPD and asthma. These formulations were prepared from spray-dried particles of curdlan and theophylline and the effect of curdlan on controlled release of the drug after oral administration was investigated [52-54]. Curdlan was also used in the preparation of sustained-released suppositories for delivery of indomethacin, prednisolone or salbutamol sulfate [55]. Hydroxyethylated curdlan gels were used for protein drug delivery using BSA as a model protein for *in vitro* studies [56].

Curdlan or derivatives of curdlan were also evaluated as part of nano- and microparticulate drug delivery systems. For example, microparticles of curdlan synthesized through cross-linking with epichlorohydrin in organic suspension media and further modified with strong or weakly acid anionic and palmitoyl hydrophobic groups were evaluated for enzyme and vaccine delivery. The results with lysozyme and tetanus anatoxin vaccine showed drug retention induced by electrostatic and hydrophobic forces, and *in vitro* release studies support the use of these curdlan microparticles as a potential controlled release system [57].

Another curdlan derivative, carboxymethylated curdlan was used to prepared self-assembling hydrogel nanoparticles. The self-assembly process is due to the presence of sulfonylurea as a hydrophobic moiety. The physical properties of the carboxymethylated curdlan nanoparticles are dependent on the degree of sulfonyurea substitution. These nanoparticles were evaluated for potential drug delivery to hepatic carcinoma cell line (HepG2) using lactobionic acid as payload drug [58].

Drug Delivery Applications with Scleroglucan

Scleroglucan (SG) is a β(1-3) branched β(1-3)-D-glucan with a rigid triple helix structure produced by fungi of the genus *Sclerotium.* It has been applied as coating in liposome formulations and in the preparation of hydrogels. Hydrogels are three-dimensional, hydrophilic, polymer networks capable of swelling in the presence of water or biological fluids. Natural polymers are advantageous over synthetic polymers in the preparation of hydrogels for drug delivery due to their low toxicity and biocompatibility. Despite structural differences, SG is capable of forming gels and shows similar drug release profile as gels obtained from other natural polymers such as the galactomannans, Guar gum and Locust bean gum [59].

The development of SG hydrogels for drug delivery can be classified into three groups of materials: (1) natural scleroglucan used in the preparation of sustained released tablets and ocular formulations, (2) oxidized and cross-linked derivatives of scleroglucan to generate materials for hydrogels sensitive to environment conditions, and (3) co-crosslinked SG or SG derivatives with other polymers (*i.e.,* gellan, alginate), or with salts (calcium chloride, borax) [60].

Natural Scleroglucan Gels

The first uses of scleroglucan for hydrogel formation demonstrated the ability of SG to form a monolithic swellable matrix with non-Fickian release properties under different active drug concentrations. The SG gels were not affected by environmental pH conditions and were suitable for the formulation of slow release non-disintegrating oral dosage forms [61-64]. The release of drugs from SG hydrogels was studied using model drugs like theophylline and showed significant release rate differences dependent on the SG content (% w/w) in the gel. The microenviromental properties of the SG gel cavities have been proposed to play a key role in the release of pharmaceutical drugs [65, 66], as well as the effect of additives, and gel erosion [67, 68].

The rheological properties (viscoelasticity, water uptake, and dimensional swelling) of SG gels vary according to the SG content, crosslinking reagents, and method of hydrogel preparation. Reinforced physically cross-linked networks have been obtained by preparation of hydrogels using freeze-thaw methods rather than room-temperature drying methods [69].

Gels from Cross-Linked Derivatives of Scleroglucan

It has been of interest to study the effects of different crosslinking agents on gel formation using chemical derivatives of SG due to the lack of pH response and low gel formation at SG concentrations below 1% (w/w) from natural SG gels. Carboxylated scleroglucan was studied for improved hydrogel formation and sustained release of drugs. Carboxylated SG formed a reversible, pH dependent sol-gel drug delivery systems [70, 71]. As with native SG hydrogels, carboxylated SG has been studied for loading/release of drugs using compounds of very different molecular weight and steric hindrance (*i.e.* theophylline, vitamin B12, and myoglobin).

Improvement in gel formation with carboxymethyl SG ionically crosslinked through the addition of $CaCl_2$ allowed for a complete transition toward a gel [72, 73]. Carboxymethyl SG hydrogel strength is related to polymer and $CaCl_2$ concentrations. Gels of carboxymethyl SG with different Ca^{+2} ratios have been used to evaluate delivery and release of non steroidal anti-inflammatory drugs (*i.e.,* diclofenac). These hydrogels have shown good biocompatibility in initial skin irritation tests, and the use of optimal carboxymethyl SG/Ca(II) ratios resulted in hydrogels that were able to release drugs with zero-order kinetics, making these hydrogels suitable for the development of topical formulations [74, 75].

Crosslinking of SG has also been investigated by carboxylation of activated CO_2 followed by alkylation with bifunctional substituents, and acylation with dicarboxylic acids for sustained/controlled release [76]; preparation of polycarboxylated derivatives of SG and 1,6-hexane dibromide [77-79], oxidation with sodium periodate to form scleraldehyde and subsequent crosslinking with hexamethylenediamine [80], cross-linking with 1,o-dicarboxylic acids having a number of carbon atoms from 4 to 8 in the chain [81], oxidation of glucopyranose side chains of SG [82], and selective C-6 halogenation to prepare SG conjugates with the antifolate drug methotrexate [83].

• Co-crosslinked SG gels. The use of additives, and co-crosslinking of SG with inorganic compounds and other polymers has been studied to evaluate effects on hydrogel formation, stability, and effect of loading/release of pharmaceutical drugs. Co-crosslinked polysaccharides of gellan and SG [84, 85] form hydrogels that show superior sustained release characteristics than obtained with the single polysaccharide hydrogels.

Carboxymethyl SG (CMSG) has also been used to prepare hydrogels in combination with alginate. The use of carboxymethyl SG improves the drug release characteristics of alginate hydrogels. Hydrogels of Alginate/CMSG are suitable as a carrier for controlled release of protein drugs in a pH responsive manner [86].

Hydrogels of SG and borax are one of the major improvements in the application of SG hydrogels for drug delivery [87]. SG forms strong physical gels by crosslinking with borate anions. SG-borax forms a channel structure mediated *via* borax ion interaction that accommodates guest molecules of different size [88]. The release profiles of different molecules have been studied by molecular dynamic simulations [89], and direct measurement of drug release from tablets containing SG/borax hydrogels with drugs of different molecular weight [90]. The advantage of SG/borate gels is that they show strong pH dependence. The gel is dissolved at a very alkaline pH due to the SG undergoing a conformation transition from a triple helix to a coil at pH 13. The gel is also dissolved in very acidic media as result of the dissociation of boric acid leading to decomplexation of the SG/borax system [91]. Alhaique *et. al.*, recently reported the use of borate as well as other ions such as aluminum and iron to prepare SG hydrogels and tablets with drugs of different steric hindrances (theophylline, vitamin B12 and myoglobin). Their results demonstrated similar drug release properties regardless of the crosslinking agent used in the preparation of the gels [92]. Additionally, the incorporation of alginate chains interspersed in SG/borax gels increased the hydrogel storage modulus by an order of magnitude and has been applied for the release of myoglobin following enteric delivery [93, 94].

Nanoparticles Coated with β(1-3)-D-Glucans

In other nanoparticle applications, curdlan was used to coat and stabilize the surface of other drug carriers (*i.e.,* liposomes) and facilitate oral delivery [95]. Curdlan was also used as a shell for solid lipid nanoparticles (SLN) prepared with cacao butter. The nanoparticles were stabilized in Tween 80 to prevent particle aggregation and curdlan gelling, and the particles were used to evaluate drug loading/release using a pyrene compound as model fluorescent probe [96]. SLN of glyceryl caprate as lipid core were coated with curdlan and stabilized in the presence of polyethylene glycol 660 hydroxystereate as a surfactant. These SLN were used to evaluate entrapment efficiency, drug loading capacity, and release behavior with doxorubicin [97].

O-palmitoylscleroglucan (PSCG) has been used to prepare PSCG-coated liposomes for oral delivery of peptide drugs. The polysaccharide coated liposomes were able to minimize the disruptive influences of gastrointestinal fluids on peptide drugs. DSC studies reveal that both SG and PSCG coated liposome surfaces with the PSCG providing stabilization by anchoring to surface [98].

Use of Soluble β(1-3)-D-Glucans as Macromolecular Drug Delivery Vehicles

Soluble β(1-3)-D-glucans have been studied for the encapsulation of macromolecular drugs such as DNA and proteins. Formation of DNA/glucan complexes allows for protection of the payload molecule and facilitates delivery to macrophage cells and other cells containing β(1-3)-D-glucan receptors. Schizophyllan (SPG) and to a lesser extent curdlan [99] have been studied for complex formation with oligonucleotides. Non-viral methods for delivery of macromolecules such as nucleic acids typically rely on the formation of electrostatically bound complexes between the anionic nucleic acids and cationic natural or synthetic polymers and lipids. The mechanism of oligonucleotide binding to neutral β(1-3)-D-glucans is different from the complex formation with charged molecules and can potentially present disadvantages regarding complex stability. However, improvements to the formulation of glucan/oligonucleotide complexes have been accomplished through synthetic modification of β(1-3)-D-glucans such as SPG. The work of Sakurai and co-workers in the past 10 years has been focused on the application of SPG and its synthetic derivatives to oligonucleotide (DNA and siRNA) delivery [99, 100], and other applications such as complexation of schizophyllan with m-carborane for drug delivery and boron neutron capture therapy [101, 102].

Schizophyllan is a β(1-3)-glucan with a single glucose side chain through a β1,6-glycosidic bond. It has a triple helix structure in water and a single chain structure (s-SPG) in DMSO. Mixtures of SPG and a single stranded oligonucleotide in DMSO can form a complex when the mixture is transferred to water. The SPG-oligonucleotide complex forms a triple helix structure with the SPG serving as a complementary sequence to the oligonucleotide. The initial work reported by Sakurai, *et. al.* proposed the use of SPG for RNA separation material,or as a novel gene carrier [103-106]. SPG has no charge, which makes it a very different gene carrier, as materials commonly used for non-viral DNA or siRNA delivery agents are cationic polymers, such as polyethyleneimine (PEI) [107]. Although SPG has no cationic charge, it effectively complexes single stranded RNA or DNA through hydrogen bonding and hydrophobic interactions. Circular dichroism (CD) spectra showed that the formation of SPG-polynucleotide complexes required a polymer length of at least 30 glucose residues. The CD results provided evidence that SPG and also lentinan can interact with polynucleotides (*i.e.,* single stranded RNA poly(C)), unlike other polysaccharides such as amylase, pullulan, dextran, and curdlan [108].

Additional studies have shown the specificity of SPG to complex certain oligonucleotides (poly (C), poly(A), poly(dA), and poly(dT)), but not poly (G), poly (U), poly(I), poly(dG), or poly(dC). This nucleotide specificity provides evidence that hydrogen bonds are essential to form the SPG-polynucleotide complex, as nucleotides such as poly(C) have an unoccupied hydrogen-bond. Solvent composition (DMSO/water) dependence also demonstrates that hydrophobic interactions are important to form these complexes. Thermodynamic data (entropy, enthalpy change) indicates that s-SPG behaves as if it were a complementary polynucleotide chain for the corresponding polynucleotide. Stoichoimetric study suggests that the complex is a triple helix consisting of two s-SPG and one poly(C) or poly(A) chains [109]. Complex stability has been found to be dependent on polynucleotide base length [110].

These studies of SPG-polynucleotide complexes have led to a better understanding of polysaccharide-polynucleotide interactions and the application of SPG to gene therapy. Although initial studies showed that the same experimental procedure to form complexes between SPG and poly(C) cannot be applied to polynucleotides like poly(U), it has been possible to vary experimental conditions to successfully complex SPG to other polynucleotides that lack the proper hydrogen bonding abilities to bind SPG. Specifically, the addition of cations has been shown to induce complexation and increase the stability of SPG/poly(U) complexes. Fluorescence polarization experiments and ^{23}Na-NMR spectroscopy indicate that the hydroxyl group in SPG and the phosphate anion in poly(U) synergistically form a specific ligand system for the cations [111]. Schizophyllan can also complex mRNA under certain pH conditions. Complexation is accelerated when the pH is changed from 13 to 7-8 [112].

Studies of SPG for Nucleic Acid Delivery

The ability of SPG to deliver nucleic acids was first demonstrated by studies in which complexes of SPG and antisense DNA (ODN) resulted in enhanced antisense effect due to protection of the SPG-ODN complex against nuclease hydrolysis [113] and binding to albumin in the culture medium [114]. Subsequent studies have focused on (1) increasing antisense oligonucleotide delivery and antisense effect by chemically modifying SPG to enhance complex stability, (2) targeted delivery of SPG-ODN complexes, (3) use of SPG to complex double stranded nucleic acid materials, and (4) use of SPG for multiple drug delivery

Enhanced SPG-Oligonucleotide Complex Stability

Stability studies of the SPG-oligonucleotide complexes revealed that oligoamines induced complex dissociation [115]. An unusual feature of the SPG-oligonucleotide complexes is that the total charge of the complex is negative. Most non-viral gene delivery carriers are made from cationic polymers and lipids that complex negatively charged nucleotides through electrostatic interactions resulting in a positively charged complex. In contrast, SPG is neutral and therefore the total charge of the SPG-ODN complex is due to the negatively charged oligonucleotide, which likely influences the effect of cationic oligoamines dissociating SPG-ODN complexes.

Chemical modification of SPG *via* a combination of oxidation with periodate and reductive amination to introduce cationic charge (amino groups, *i.e.,* 2-aminoethanol) provided a material that demonstrated enhanced stability of the poly(C) complexes and other nucleotides compared to the complexes prepared with unmodified SPG [116]. Additionally, modified SPG induced complexation with poly(U), which does not form a complex with unmodified SPG [117]. These improvements in complex stability were obtained with low levels of chemical modification as SPG derivatization with 2-aminoethanol was done only at 2.4%.

Other modifications of SPG include the use of oligoamine conjugates of SPG that have shown *in vitro* transfection efficiencies superior to that of PEI and with lower cytotoxicity [118]. Also, the use of polyethyelene glycol (PEG) chains were used to synthesize a PEG-SPG derivative in which the PEG protects the AS ODN from lysosomal degradation [119].

Targeted Delivery of SPG-ODN Complexes

Chemical modification of SPG by reductive amination of the side chain allows attachment of ligands to enhance cellular uptake. Octa-arginine (R8) or arginine-glycine-aspartic acid (RGD) peptide derivatized SPG were used to form complexes with AS ODN resulting in enhanced antisense effect due to targeted endocytosis through interactions of the R8 or RGD peptides with integrin receptors [120]. Other materials that have been used to increase cellular uptake are spermine and cholesterol. Delivery of CpG DNA with spermine, cholesterol, RGD, or R8 modified SPG showed increased cytokine secretion by macrophages (5-10 times higher than naked CpG DNA, and 100 times higher than background) [121-123].

Similarly, reductive amination of SPG has been carried out to incorporate ligands to target specific cell surface receptors. Examples of these modifications include lactose-appended SPG or galactose-SPG to effectively mediate gene transfection into hepatocytes *via* asialoglycoprotein receptor [124, 125], folate-SPG that showed specific affinity toward folate binding protein which mediated effective antisense activity in cancer cells [126], and transferrin-SPG (Tf-SPG) [127].

In addition to demonstrating that chemical modifications of SPG enhanced cellular uptake, additional work has shown it possible to increase DNA nuclear uptake. Nuclear transport was optimized by reducing the molecular weight of SPG to 25 kDa. Low molecular weight SPG allowed for efficient transport of SPG/AS ODN complexes, although it decreased stability. Importin-B, a nuclear transport protein, was attached to the side chain of 25 kDa SPG *via* a streptavidin-biotin interaction. Telomerase activity was suppressed in Jurkat cells when using Importin-B-SPG/AS ODN complexes indicating that the protein enhanced nuclear transport [128].

SPG for Gene Delivery of Double Stranded DNA

All preliminary work with SPG showed it capable of complexing only single-stranded oligonucleotides [105, 106, 108, 115, 116]. To develop SPG for double-stranded DNA (dsDNA) or plasmid DNA transfection a minimal-size gene having a poly(dA)80 loop on both ends formed a complex with SPG. An siRNA-coding dsDNA with the poly(dA) loop was complexed with Tat-attached SPG to silence luciferase expression. Tat-SPG complex induced much less cell death than PEI [129].

Another approach to use SPG to complex DNA heterosequences without attachment of homopolynucleotide tails is to form polycation/DNA/SPG ternary complexes. Polycation/DNA complexes are formed and then encapsulated by the SPG hydrophobic domain. The ternary complexes showed high dectin-1 mediated uptake efficiency in macrophage-like cells (J774.A1). IL-12 secretion was enhanced when CpG-DNA was added as the ternary complex [130].

Multiple Drug Delivery

SPG has been used deliver the model antigen ovalbumin (x OVA) and an adjuvant (CpG DNA). OVA-SPG conjugates were used to complex CpG DNA enabling both CpG DNA and antigen delivery to the same cell. The macrophage cell line J774.A1 showed increased OVA ingestion compared to OVA alone suggesting the dectin-1 receptor is involved in OVA-SPG complex ingestion. Furthermore, the OVA and DNA were co-localized in the same vesicles indicating that both DNA and the antigen were co-ingested [131]. CpG DNA delivery *in vivo* (i.p. injection) showed a large amount of IL-12 production compared to the uncomplexed materials [132].

β(1-3)-D-Glucan Particles as a Drug Delivery System

β(1-3)-D-Glucan derived from baker's yeast can be processed into hollow, highly porous microparticles. Fig. **1** depicts photomicrographs of a whole yeast cell (Fig. **1a**) and a diagrammatic cross-section of a cell highlighting the cell wall structure (Fig. **1b**). The cell wall is rich in β(1-3)-glucan fibrils and by a series of chemical extractions it is possible to remove cytoplasmic and membrane constituents yielding empty spherical particles with a diameter of 2-4 μm. Depending on the chemical conditions and yeast source it is possible to prepare glucan particles (GP) with different ratios of β(1-3)-glucan, mannan, and chitin. The different classes of GPs and their percentage composition are summarized in Table **2**.

The GPs are efficiently phagocytosed by cells bearing β(1-3)-D-glucan receptors making them suitable for targeting to macrophage and dendritic cells present in the immune system. These particles can be administered by oral, topical, inhalation and parenteral routes. The mechanism of oral glucan particle absorption relies largely on M cells within Peyers patches facilitating transportation of β(1-3)-glucans across the intestinal cell wall into the gut associated lymphatic tissue (GALT) where they interact with macrophages to activate immune function [133]. Following particle uptake by intestinal macrophages cellular trafficking transports GPs to different tissues and organs (bone, lung, liver and spleen), and sites of inflammatory pathology [10, 134].

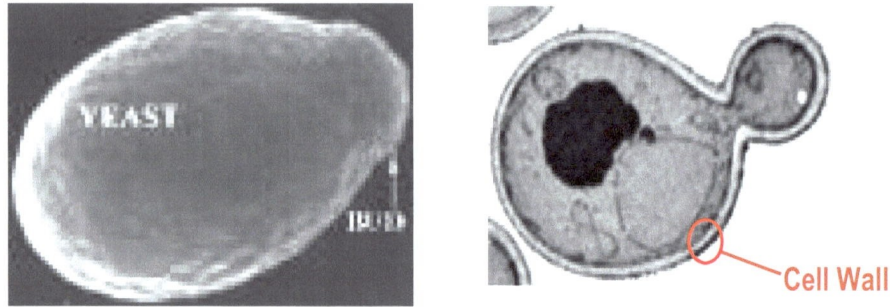

Baker's Yeast Cell Wall

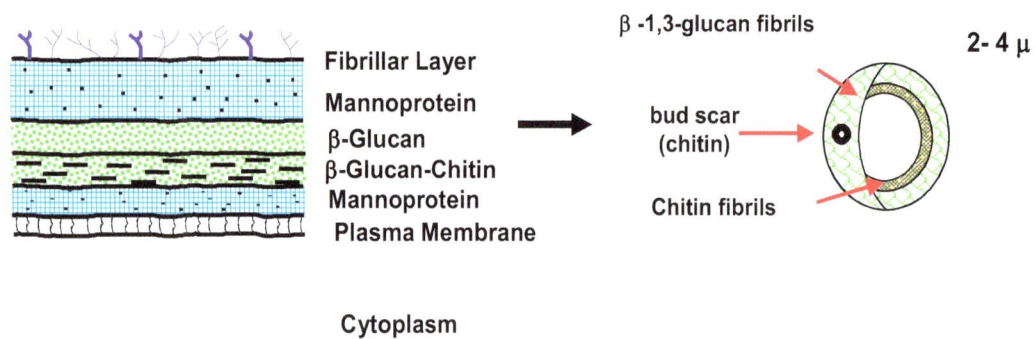

Figure 1: (a) Scanning Electron Micrograph of a yeast cell (b) Transmission electron micrograph showing a cross section of a yeast cell (c) Schematic representation of the process to prepare glucan particles following chemical extraction of the components of Baker's yeast.

Table 2: Composition of different glucan particles (From Reference [136])

Composition	Particle type		
	Glucan particle (GP)	Glucan mannan particle (GMP)	Yeast glucan chitin
Glucan	80%	40%	40-50%
Mannan	<1%	40%	0%
Chitin	2-4%	2-4%	40-50%

Research on βglucan microparticles dates back to the early 1990's when Alpha-Beta developed glucan carbohydrate microcapsules (Adjuvax) for targeted antigen and drug delivery [136-138]. Adjuvax was produced as a highly purified carbohydrate vaccine microcapsule combining three key adjuvant properties: (1) macrophage targeting, (2) macrophage activation, and (3) sustained release of antigens. Adjuvax-ligand complexes were formed by chemical crosslinking of the ligand within the glucan microcapsules (Fig. **2**). The advantages of the Adjuvax-antigen conjugates was demonstrated by *in vivo* macrophage targeting of antigens such as OVA with 1,000 fold response over antigen only controls and lack of adverse side-effects [137].

Disadvantages of chemical crosslinking the target molecule (*i.e.,* peptide, protein) with the particles include the length and yield of synthetic procedures, limited payload binding capacity, and the effect of covalent cross linking on release rate and antigenicity. The next step in the development of glucan particles as efficient drug delivery vehicle was the design of synthetic strategies for encapsulation of payload molecules within the glucan particles as electrostatically bound polymer complexes (polyplexes) between payload molecules and trapping polymers of opposite charge [139, 140].

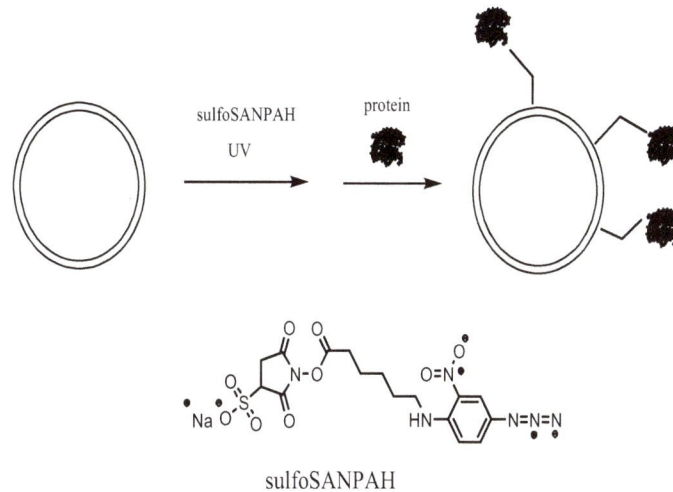

Figure 2: Schematic representation of chemical cross-linking of payload molecules to glucan particles.

Synthesis of encapsulated polyplexes inside glucan particles consists of two steps: (1) polyplex core formation and (2) Layer-by-Layer (LbL) assembly of electrostatically bound materials. This synthetic strategy allows for the formation of three classes of GP encapsulated formulations (Fig. **3**). A polyelectrolyte core material is absorbed into the hollow, porous glucan particles by incubation of the particles in a solution of the core material (*i.e.,* tRNA, DNA, proteins). To prevent core elution from the particles a trapping polymer of opposite charge to the core is added to form a polyplex large enough to prevent leakage from the GPs. Core formulations can contain the payload molecule as part of the core (*i.e.,* siRNA, DNA, proteins), or can be made of a non-toxic, inert material (*i.e.,* synthetic or natural biopolymers, inorganic complexes, or nanoparticles) and this core can serve as surface onto which subsequently added layers of polyelectrolyte materials are adsorbed using LbL synthesis methods to yield "layered formulations". A "core+layered formulation" contains the payload molecule in both locations. The position of the payload molecule in the formulation is selected based on drug loading levels required and effect on release (generally payload molecules in the core are released more slowly than from layered formulations).

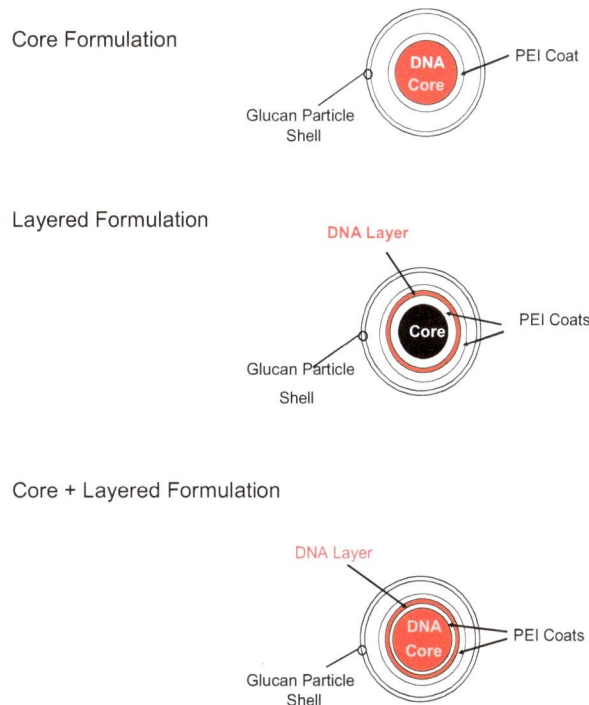

Figure 3: Synthetic strategies to encapsulate payload molecules inside GPs

These methods of encapsulating payload molecules within GP provides the advantages of rapid and efficient loading, the ability to multiplex load more than one payload molecule, and the ability to control loading and formulation composition to optimize payload release. Glucan particles have been used to prepare encapsulated polyplexes for DNA, siRNA, protein and nanoparticles, and other encapsulation methods have been developed to use glucan particles for the delivery of small drug molecules. These GP encapsulated formulations can be used to target and deliver drugs to macrophage and dendritic cells. Additional chemical modification of the glucan particle surface allows for targeted delivery to cells that do not express β-glucan receptors on their surface and, thus are not capable efficiently ingesting native glucan particles.

1. DNA delivery. Core and LbL synthesis of DNA encapsulated polyplexes has been developed using a model system with a plasmid expressing green fluorescent protein (GFP). The DNA polyplexes were built using either a GFP plasmid DNA core, or layered onto a tRNA/PEI core where the plasmid was bound to the cores inside the particles between layers of cationic PEI, or other trapping polymers (Fig. **4**). The formation of cores and DNA encapsulated polyplexes was demonstrated using fluorescently labeled PEI and nucleic acids to quantify the amount of trapped materials and optimize loading conditions in each step of the synthesis [141]. By preparing formulations with fluorescent tRNA or fluorescent PEI it was possible to quantify adsorption of each layer using a fluorescence based assay and fluorescence assisted cell sorting (FACS) measurements. Formulations containing two fluorescently labeled materials (*i.e.,* rhodamine labeled PEI, and fluorescein labeled tRNA) were used to show co-localization of both polyplex components inside the particles using fluorescence microscopy.

Particles loaded with a GFP expressing plasmid using optimal loading conditions determined from fluorescent DNA delivery studies were evaluated for transfection activity using a fibroblast murine cell line NIH3T3-D1. The GP encapsulated DNA formulations showed no significant cytotoxicity or adverse effect on cell growth, and high levels of transfection were obtained with less than 100 ng DNA/1×10^6 particles/1×10^5 cells. The level of transfection at any DNA concentration was 3-5 times higher than transfection obtained with naked DNA/PEI polyplexes [141].

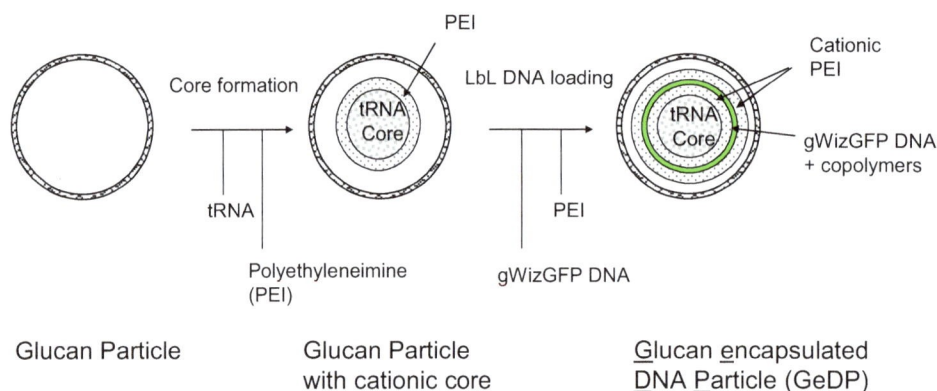

Figure 4: Layer-by-layer (LbL) synthesis of DNA encapsulated polyplexes

DNA transfection has been optimized by incorporating compounds in the layered formulations that can have a stimulatory effect on (1) DNA/PEI polyplex defoliation and release from the particles, (2) DNA endosomal release, and (3) DNA nuclear uptake. Preliminary studies have shown enhancement of GFP DNA transfection on NIH3T3-D1 cells using formulations containing compounds such as Endo-Porter (a endosomal release peptide), or penetratin (a nuclear localization sequence (NLS) peptide) [142].

2. siRNA delivery. The preparation of glucan particles for siRNA delivery followed a similar approach developed for DNA formulations. A co-delivery system in which GFP DNA and GFP siRNA was developed to demonstrate the ability of GPs to efficiently deliver siRNA [143]. Glucan particles containing GFP DNA and a scrambled siRNA showed high levels of GFP expression by NIH3T3-D1 cells. GP formulations containing GFP DNA and GFP siRNA showed markedly lower GFP expression levels with the level of GFP silencing dependent on siRNA concentration. GP siRNA encapsulation provides protection of siRNA from nuclease degradation and targeted delivery *via* dectin-1 and CR3 receptors resulting in gene silencing with low doses of siRNA.

A significant advance in the development of glucan particles for siRNA delivery was recently reported by Aouadi, *et al.* [134]. In these studies, glucan encapsulated siRNA particles (GeRPs) were used as an efficient oral siRNA delivery vehicle to silence genes in mouse macrophages *in vitro* and *in vivo*. Oral gavage of mice with GeRPs containing a mitogen-activated protein kinase (Map4K4) siRNA effectively silenced Map4K4 and TNFα (Map4K4 controls TNFα expression in macrophages) in peritoneal exudate cells, and splenic and liver macrophages compared to scrambled siRNA controls. Treatment with Map4K4 GeRPs protected mice against lipopolysaccharide-induced lethality by inhibiting TNFα production. The significance of the study is the achievement of successful siRNA delivery following oral administration of as low as 20 µg siRNA/kg.

3. Protein delivery. The Adjuvax technology was used for delivery of chemically crosslinked proteins or chemically conjugated to the glucan particles. Three proteins of different molecular weight (cytochrome-C, cyt C, Mw = 14000 Da; bovine serum albumin, BSA, Mw = 67000 Da, and alcohol dehydrogenase ADH, Mw =150000 Da) were loaded into glucan particles and prepared as dried mixtures of GP/protein. Release studies demonstrated that the release rate of the protein was dependent on the molecular weight of the protein. In order to increase retention a mutant strain of yeast, *Saccharomyces cerevisiae R4*, was used to prepare particles with higher degree of β(1-6) branching that provided lower permeability and longer retention times for BSA compared to particles derived from Baker's yeast [137]. Uncomplexed protein payload release rates occurred rapidly requiring either chemically crosslinking proteins to each other, or to the glucan shell to stably associate the protein with the GPs. Glucan particles with chemically conjugated proteins were obtained by reaction of a molecule like cytochrome-C with the heterobifunctional cross-linking reagent sulfosuccinimidyl 6-(4'-azido-2'-nitrophenylamino)hexanoate (sulfo-SANPAH) [136].

Proteins can also be encapsulated within glucan particles using a polyplex synthesis strategy. Glucan particles containing an Ovalbumin/ polyethylenimine (OVA/PEI) core were used to transfect GFP DNA in the model NIH3T3-D1 system [144]. The studies with model proteins, such as BSA and OVA have been done to develop methods for the co-delivery of antigens, DNA, siRNA and PAMP adjuvants for vaccine applications.

In another variation of the GP vaccine delivery technology recombinant yeast expressing protein antigens as virus-like particle fusions or fibril forming scaffold protein fusion proteins are processed to expose cell wall β(1-3)-D-glucan providing targeted antigen delivery [145].

4. Nanoparticle encapsulation. The original synthetic procedures for encapsulation of payloads within glucan particles were designed to trap soluble molecules by formation of encapsulated polyplexes that are sufficiently large to prevent their escape from the glucan particles. Recent advances in the glucan particle technology have led to the development of methods to encapsulate preformed nanoparticles. The limitation in this procedure is that nanoparticles should have a diameter of less than 30 nm to facilitate efficient passage through the porous glucan shell during the loading process. After nanoparticle loading larger aggregates are formed by crosslinking with polymers (Fig. **5**). Using this method we have been able to encapsulate fluorescently labeled 20 nm polystyrene, 10 nm magnetic, and 25 nm quantum dot nanoparticles (Fig. **6**). The advantages of nanoparticle encapsulation are: 1) it extends the applicability of the glucan particle delivery system by allowing encapsulation of materials that cannot be prepared *in situ* as the synthetic conditions are not compatible with glucan particles and 2) it allows incorporation of nanoparticles with an intrinsic property, such as magnetic nanoparticles thus increasing the usefulness of the particles, as the same formulation could be used for drug (siRNA, DNA, protein) delivery and the magnetic properties employed for purification of cells, or for imaging applications.

Figure 5: Schematic diagram showing loading of nanoparticles within glucan particles

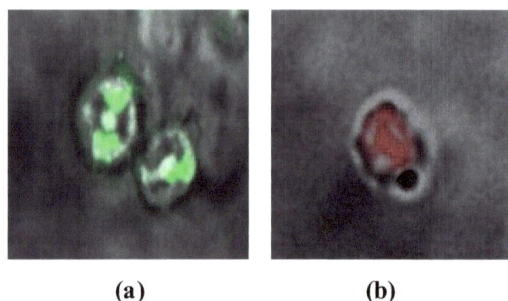

(a)   (b)

Figure 6: Fluorescent micrographs showing encapsulation of (a) quantum dots, and (b) magnetic nanoparticles crosslinked with fluorescent PEI

5. Small drug delivery. The synthetic methods used to encapsulate payload macromolecules, such as proteins and DNA are based on non-covalent electrostatic interactions, and this approach could be applicable to certain charged small drug molecules. However, the majority of small drug molecules are neutral or insoluble in water and therefore unlikely to be effectively trapped within glucan particles using an LbL encapsulation method as described previously. However, we have successfully demonstrated the use of glucan particles for delivery of small drug molecules using different encapsulation protocols. Two examples of these molecules are rifampicin and terpenes.

Rifampicin is an antibiotic that is used as a part of a drug-cocktail to treat tuberculosis (TB). TB infects and replicates inside macrophages. The ability of glucan particles to target macrophage cells makes it an alternative method to effectively deliver high doses of rifampicin to TB infected alveolar macrophages. Rifampicin was loaded in dry glucan particles by adsorption of a rifampicin solution in 0.2 N HCl. A rapid change in pH with addition of a pH 8 buffer caused precipitation of rifampicin inside the particles. In order to prevent or slow down release of the encapsulated drug, the rifampicin loaded particles were sealed with a polymer capable of forming gels to fill the pores of the glucan particles. Fig. 7 shows the rifampicin release profile of an unplugged sample and a rifampicin loaded sample that had been sealed with an alginate gel obtained from crosslinking of alginate in the presence of calcium chloride (Fig. 7).

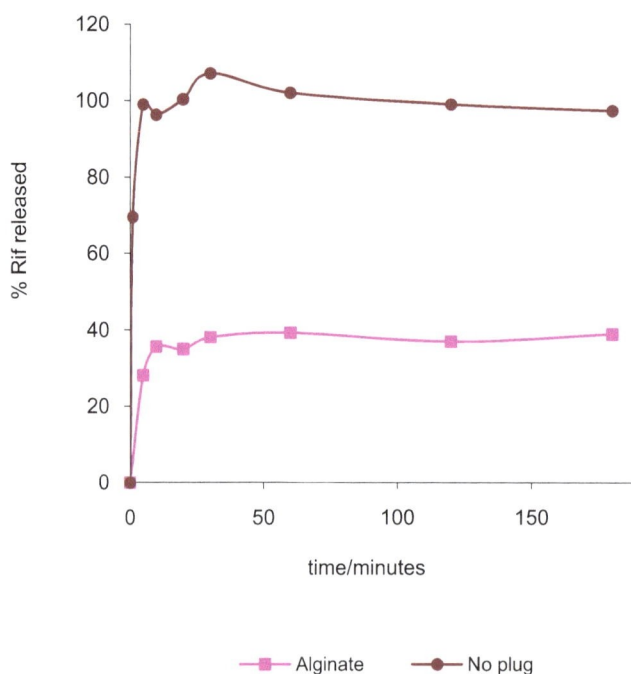

Figure 7: Effect of alginate plug on release rate of rifampicin from GP particles

Glucan particles have been used for encapsulation of terpenes for agro-chemical applications [146-149]. Glucan particle encapsulated terpenes provide for a water-suspendable formulation and moisture-activated sustained terpene release that is effective against a broad range of fungal plant pathogens. Different natural terpene compounds (eugenol, geraniol, thymol, and citral) and mixtures of these materials were homogenized in a surfactant mixture of 0.01% Tween 80/0.01% xanthan gum and used to prepare GP encapsulated terpene formulations. The GP encapsulated terpenes are being developed as Botrytis, Downy and Powdery mildew treatments in vineyards and have shown comparable fungicide activity to synthetic fungicides.

6. Targeted delivery with chemically modified GPs. The native glucan particles are phagocytosed by macrophage and other cell types bearing β(1-3)-D-glucan receptors. To target glucan particles to cells that do not bear glucan receptors GPs have been chemically derivatized to incorporate molecules onto their surface that will be recognized by specific cell surface receptors. Incorporation of ligands can be accomplished by direct linkage to the GP using a polyethylene glycol (PEG) or other non-reactive spacers, or by derivatization of the glucan particles with molecules that can be used as a universal surface for attachment of ligands based on specific chemical interactions, *i.e.,* glucan particles derivatized with cyclodextrin interacting ligands bearing adamantane groups *via* host-guest interaction between cyclodextrin and adamantane, or glucan particles derivatized with biotin for binding of ligands through biotin-avidin or streptavidin interactions.

These GP surface derivatization approaches have been used to produce galactose-modified glucan particles for hepatocyte targeting *via* the asialoglycoprotein receptor (ASGPr), which recognizes galactose trimers as ligands [150]. Chemical modification of glucan particles by a reductive amination approach to couple a PEG-galactose and subsequent *in vitro* studies with hepatocyte (HepG2) cell line have demonstrated the ability to modify particles for targeted delivery to cells that do not show uptake of native GPs (Fig. **8**).

(a)   (b)

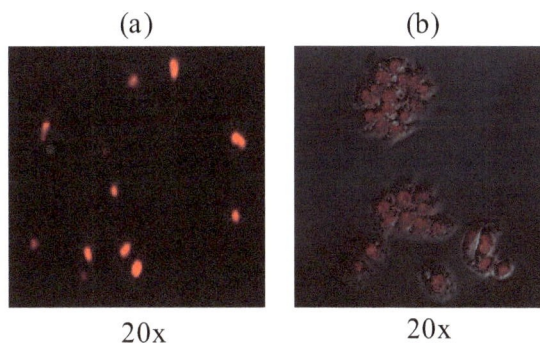

20x   20x

Figure 8: (a) Fluorescent micrograph of GP-PEG-galactose particles containing a tRNA/PEI core with a layer of Dy547-fluorescently labeled siRNA (b) Micrograph showing avid GP-Gal-Dy547-siRNA uptake by hepatocyte cells (HepG2)

CONCLUSIONS

The immunomodulatory and physical properties of β(1-3)-D-glucans have been studied for use in a wide range of human applications and β(1-3)-D-glucans are found, for example in foods as thickeners or soluble dietary fiber, in topically applied gels and wound dressings, in oral nutraceutical preparations, and injectable therapeutics. The immunomodulatory properties of β(1-3)-D-glucans are a function of receptor-mediated interactions targeting specific cells in the innate immune system, such as monocytes, macrophages, neutrophils, NK T-cells and dendritic cells. β(1-3)-D-glucans also possess unique physical properties allowing them to be produced as a solution, gel, film or particle suspension and when used alone or in combination with other components are useful for drug delivery applications. In some applications the β(1-3)-D-glucan serves as both a carrier and the active, in other applications β(1-3)-D-glucan formulations have been shown useful to deliver a diverse range of drugs including; 1) small molecules, 2) peptides and proteins, 3) anti-sense oligonucleotides, siRNAs and DNAs, and 4) particulate drugs. Exciting future uses of β(1-3)-D-glucans for drug delivery will capitalize on the intersection of the unique immunomodulatory and physical properties of this class of polysaccharides to develop drug formulations targeting innate immune cells, tissues rich in these cells and sites of inflammatory pathology.

ACKNOWLEDGMENTS

We acknowledge the contributions of the many collaborators cited in published and unpublished work, and the financial support from NIH, Juvenile Diabetes Research Foundation, Gates Foundation, Mass Life Sciences, University of Massachusetts Medical School and Commonwealth Medicine.

REFERENCES

[1]   Novak M, Vetvicka V. β-glucans, history, and the present: immunomodulatory aspects and mechanisms of action. J Immunotoxicol 2008; 5: 47-57.

[2]   Smelcerovic A, Knezevic-Jugovic Z, Petronijevic Z. Microbial polysaccharides and their derivatives as current and prospective pharmaceuticals. Curr Pharm Res 2008; 14: 3168-3195.

[3]   Vetvicka V, Thornton BP, Ross GD. Soluble βglucan polysaccharide binding to the lectin site of neutrophil or natural killer cell complement receptor type 3 (CD11b/CD18) generates a primed state of the receptor capable of mediating cytotoxicity of iC3b-opsonized target cells. J Clin Invest 1996; 98: 50-61.

[4]   Thornton BP, Vetvicka V, Pitman M, Goldman RC, Ross GD. Analysis of the sugar specificity and molecular location of the βglucan-binding lectin site of complement receptor type 3 (CD11b/CD18). J Immunol 1996; 156: 1235-1246.

[5]   Brown GD, Gordon S. Immune recognition: a new receptor for β-glucans. Nature 2001; 413: 36-7.

[6]   Taylor PR, Brown GD, Reid DM, *et al.* The β-glucan receptor, dectin-1, is predominantly expressed on the surface cells of the monocyte/macrophage and neutrophil lineages. J Immunol 2002; 169: 3876.

[7]   Goodridge H, Wolf AJ, Underhill DM. β-glucan recognition by the innate immune system. Immunol Rev 2009; 230: 38-50.

[8]   Miura NN, Ohno N, Aketagawa J, Tamura H, Tanaka S, Yadomae T. Blood clearance of (1→3)-β-D-glucan in MRL lpr/lpr mice. FEMS Immunol Med Microbiol 1996; 13: 51-57.

[9]   Chihara G. Recent progress in immunopharmacology and therapeutic effects of polysaccharides. Dev Biol Standard 1992; 77: 191-197.

[10]   Hong F, Yan J, Baran JT, *et al.* Mechanism by Which Orally Administered β(1-3)-Glucans Enhance the Tumoricidal Activity of Antitumor Monoclonal Antibodies in Murine Tumor Models. J Immunol 2004; 173: 797-806.

[11]   Vetvicka V, Dvorak B, Vetvickova J, *et al.* Orally administered marine β(1-3)-D-glucan Phycarine stimulates both humoral and cellular immunity. International J Biol Macromol 2007; 40: 291-298.

[12]   Mucksova J, Babicek K, Pospisil M. Particulate β(1-3)-D-glucan carboxymethylglucan and sulfoethyl-glucan--influence of their oral or intraperitoneal administration on immunological respondence of mice. Fol Microbiol 2001; 46: 559-563.

[13]   Castro GR, Panilaitis B, Bora E, Kaplan DL. Controlled Release Biopolymers for Enhancing the Immune Response. Mol Pharmaceutics 2007; 4: 33-46.

[14]   Keogh GF, Cooper GJ, Mulvey TB, *et al.* Randomized controlled crossover study of the effect of a highly β-glucan-enriched barley on cardiovascular disease risk factors in mildly hypercholesterolemic men. Am J Clin Nutr 2003; 78: 711-718.

[15]   Otaka K. Functional oligo saccharide and its new aspect as immune modulation. J Biol Macromol 2006; 6: 3-9.

[16]   Dellinger EP, Babineau TJ, Bleicher P, *et al.* Effect of PGG-glucan on the rate of serious postoperative infection or death observed after high-risk gastrointestinal operations. Betafectin Gastrointestinal Study Group. Arch Surgery 1999; 134: 977-983.

[17]   Babineau TJ, Hackford A, Kenler A, *et al.* A phase II multicenter, double-blind, randomized, placebo-controlled study of three dosages of an immunomodulator (PGG-glucan) in high-risk surgical patients. Arch Surgery 1994; 129: 1204-1210.

[18]   Babineau TJ, Marcello P, Swails W, Kenler A, Bistrian B, Forse RA. Randomized phase I/II trial of a macrophage-specific immunomodulator (PGG-glucan) in high-risk surgical patients. Ann Surg 1994; 220: 601-609.

[19]   Vetvicka V, Terayama K, Mandeville R, Brousseau P, Kournikakis B, Ostroff G. Pilot Study: Orally-Administered Yeast β(1-3)-glucan Prophylactically Protects Against Anthrax Infection and Cancer in Mice. J American Nutraceutical Assoc 2002; 5: 5-9.

[20]   DiLuzio NR, Williams DL, McNamee RB, Malshet VG. Comparative evaluation of the tumor inhibitory and antibacterial activity of solubilized and particulate glucan. Recent Results Cancer Res 1980; 75: 165-172.

[21]   Mansell PW, Ichinose H, Reed RJ, Krementz ET, McNamee R, DiLuzio NR. Macrophage-mediated destruction of human malignant cells *in vivo.* J Nat Cancer Inst 1975; 54: 571-580.

[22]   Morikawa K, Takeda R, Yamazaki M, Mizuno D. Induction of tumoricidal activity of polymorphonuclear leukocytes by a linear β(1-3)-D-glucan and other immunomodulators in murine cells. Cancer Res 1985; 45: 1496-1501.

[23]   Maeda YY, Chihara G. Lentinan and other antitumoral polysaccharides. Immunomodulatory Agents Plants 1999: 203-221.

[24] Wakui A, Kasai M, Konno K, *et al.* Randomized study of lentinan on patients with advance gastric and colorectal cancer. Tohoku Lentinan Study Group. Cancer & Chemotherapy 1986; 13: 1050-1059.

[25] Shimizu K, Watanabe S, Watanabe S, Matsuda K, Suga T, Nakazawa S, *et al.* Efficacy of oral administered superfine dispersed lentinan for advanced pancreatic cancer. Hepato-Gastroenterology 2009; 56: 240-244.

[26] Hazama S, Watanabe S, Ohashi M, *et al.* Efficacy of orally administered superfine dispersed lentinan β(1-3)-glucan) for the treatment of advanced colorectal cancer. Anticancer Res 2009; 29: 2611-2618.

[27] Hanaue H, Tokuda Y, Machimura T, *et al.* Effects of oral lentinan on T-cell subsets in peripheral venous blood. ClinTherap 1989; 11: 614-622.

[28] Rice PJ, Adams EL, Ozment-Skelton T, *et al.* Oral delivery and gastrointestinal absorption of soluble glucans stimulate increased resistance to infectious challenge. J Pharmacol Exp Ther 2005; 314: 1079-1086.

[29] Suzuki I, Hashimoto K, Ohno N, Tanaka H, Yadomae T. Immunomodulation by orally administered β-glucan in mice. Int J Immunopharmacol 1989; 11: 761-769.

[30] Cheung NK, Modak S, Vickers A, B K. Orally administered β-glucans enhance anti-tumor effects of monoclonal antibodies. Cancer Immunol Immunotherap 2002; 51: 557-564.

[31] Sveinbjornsson B, Rushfeldt C, Seljelid R, Smedsrod B. Inhibition of establishment and growth of mouse liver metastases after treatment with interferon gamma and β(1-3)-D-glucan. Hepatology 1998; 27: 1241-1248.

[32] Cui SW, Wang Q. Cell wall polysaccharides in cereals: chemical structures and functional properties. Structural Chem 2009; 20: 291-297.

[33] Jamas S, Easson DD, Jr., Ostroff GR, Onderdonk AB. PGG-glucans. A novel class of macrophage-activating immunomodulators. ACS Symp Ser 1991; 469: 44-51.

[34] Demir G, Klein HO, Mandel-Molinas N, Tuzuner N. Beta glucan induces proliferation and activation of monocytes in peripheral blood of patients with advanced breast cancer. Int Immunopharmacol 2006; 7: 113-116.

[35] Kimura Y, Sumiyoshi M, Suzuki T, Sakanaka M. Antitumor and antimetastatic activity of a novel water-soluble low molecular weight β-1, 3-D-glucan (branch β-1,6) isolated from *Aureobasidium pullulans* 1A1 strain black yeast. Anticancer Res 2006; 26: 4131-4142.

[36] Tichy E, Vitkova Z, Cupkova B. Effect of β-(1,3)-glucan on rheological properties and stability of topical formulations. Pharmazie 2006; 61: 1050-1051.

[37] Patchen ML, MacVittie TJ. Dose-dependent responses of murin pluripotent stem cells and myeloid and erythroid progenitor cells following administration of the immunomodulating agent glucan. Immunopharmacology 1983; 5: 303-313.

[38] Patchen ML, MacVittie TJ. Comparative effects of soluble and particulate glucans on survival in irradiated mice. J Biol Resp Mod 1986; 5: 45-60.

[39] Patchen ML, Vaudrain T, Correira H, Martin T, Reese D. *In vitro* and *in vivo* hematopoietic activities of Betafectin PGG-glucan. Exp Hematol 1998; 26: 1247-1254.

[40] Kirmaz C, Bayrak P, Yilmaz O, Yuksel H. Effects of glucan treatment on the Th1/Th2 balance in patients with allergic rhinitis: a double-blind placebo-controlled study. Eur Cytokine Network 2005; 16: 128-134.

[41] Onderdonk AB, Cisneros RL, Hinkson P, Ostroff G. Anti-infective effect of poly-β(1-6)-glucotriosyl-β(1-3)-glucopyranose glucan *in vivo*. Infection Immunity 1992; 60: 1642-1647.

[42] Kernodle DS, Gates H, Kaiser AB. Prophylactic anti-infective activity of poly-β(1-6)-D-glucopyranosyl-β(1-3)D-glucopyranose glucan in a guinea pig model of straphylococal wound infection. Antimicrob Agents Chemother 1998; 42: 545-549.

[43] Sener G, Toklu H, Ercan F, Erkanli G. Protective effect of β-glucan against oxidative organ injury in a rat model of sepsis. Int Immunopharmacol 2005; 5: 1387-1396.

[44] Dongowski G, Huth M, Gebhardt E, Flamme W. Dietary fiber-rich barley products beneficially affect the intestinal tract of rats. J Nutr 2002; 132: 3704-3714.

[45] Battilana P, Ornstein K, Minehira K, *et al.* Mechanisms of action of β-glucan in postprandial glucose metabolism in healthy men. Eur J Clin Nutr 2001; 55: 327-333.

[46] Tsukada C, Yokoyama H, Miyaji C, Ishimoto Y, Kawamura H, Abo T. Immunopotentiation of intraepithelial lymphocytes in the intestine by oral administration of β-glucan. Cell Immunol 2003; 221: 1-5.

[47] Yao P, Li J, Yu S, Jiang M. Nanogels containing core-shell protein/polysaccharide complexes for delivery of drugs and nutrients. CN 2007-101058649. 2007 Oct.

[48] Sun K, Dou H, Lu R. Microcapsules and nanocapsules made from polymers for drug delivery. CN 2007-101053810. 2007 Oct.

[49] Yang RK, Fuisz RC. Glucan based film delivery systems. WO 2003030881. 2003 Apr.

[50]   Roeding J. β-(1,3)-β-(1,4)-Glucan as a carrier for chemical substances. WO 2006015627. 2006 Feb.

[51]   Schwartz YS, Dushkin MI, Vavilin VA, *et al.* Novel conjugate of moxifloxacin and carboxymethylated glucan with enhanced activity against Mycobacterium tuberculosis. Antimicrob Agents Chemother 2006; 50: 1982-1988.

[52]   Kanke M, Koda K, Koda Y, Katayama H, Nakayama A. Preparation and *in vitro* drug release evaluation of curdlan tablets. Drug Delivery Syst 1992; 7: 135-40.

[53]   Kanke M, Koda K, Koda Y, Katayama H. Application of curdlan to controlled drug delivery. I. The preparation and evaluation of theophylline-containing curdlan tablets. Pharm Res 1992; 9: 414-18.

[54]   Kanke M, Katayama H, Nakamura M. Application of curdlan to controlled drug delivery. II. *In vitro* and *in vivo* drug release studies of theophylline-containing curdlan tablets. Biol & Pharmaceut Bull 1995; 18: 1104-1108.

[55]   Kanke M, Tanabe E, Katayama H, Koda Y, Yoshitomi H. Application of curdlan to controlled drug delivery. III. Drug release from sustained-released suppositories *in vitro*. Biol & Pharmaceut Bull 1995; 18: 1154-1158.

[56]   Kim BS, Jung ID, Kim JS, Lee J-H, Lee IY, Lee KB. Curdlan gels as protein drug delivery vehicles. Biotechnol Lett 2000; 22: 1127-1130.

[57]   Mocanu G, Mihai D, Moscovici M, Picton L, LeCerf D. Curdlan microspheres. Synthesis, characterization and interaction with proteins (enzymes, vaccines). Int J Biol Macromol 2009; 44: 215-221.

[58]   Na K, Park KH, Kim SW, Bae YH. Self-assembled hydrogel nanoparticles from curdlan derivatives: characterization, anti-cancer drug release and interaction with a hepatoma cell line (HepG2). J Cont Rel 2000; 69: 225-236.

[59]   Coviello T, Alhaique F, Dorigo A, Matricardi P, Grassi M. Two galactomannans and scleroglucan as matrices for drug delivery: preparation and release studies. Eur J Pharm Biopharm 2007; 66: 200-209.

[60]   Coviello T, Palleschi A, Grassi M, Matricardi P, Bocchinfuso G, Alhaique F. Scleroglucan: A versatile polysaccharide for modified drug delivery. Molecules 2005; 10: 6-33.

[61]   Colombo I, Grassi M, Lapasin R, Pricl S. Determination of the drug diffusion coefficient in swollen hydrogel polymeric matrixes by means of the inverse sectioning method. J Cont Rel 1997; 47: 305-314.

[62]   Grassi M, Lapasin R, Pricl S, Colombo I. Apparent non-Fickian release from a scleroglucan gel matrix. Chem Eng Commun 1997; 155: 89-112.

[63]   Alhaique F, Carafa M, Coviello T, Murtas E, Riccieri FM, Santucci E. Release from a polysaccharide matrix: effect of the molecular weight of the drug. Acta Technol Legis Med 1993; 4: 21-30.

[64]   Alhaique F, Carafa M, Riccieri FM, Santucci E, Touitou E. Studies on the release behavior of a polysaccharide matrix. Pharmazie 1993; 48: 432-6.

[65]   Francois NJ, Rojas AM, Daraio ME, Bernik DL. Dynamic rheological measurements and drug release kinetics in swollen scleroglucan matrices. J Cont Rel 2003; 90: 355-362.

[66]   Daraio ME, Francois N, Bernik DL. Correlation Between Gel Structural Properties and Drug Release Pattern in Scleroglucan Matrices. Drug Deliv 2003; 10: 79-85.

[67]   Alhaique F, Beltrami E, Riccieri FM, Santucci E, Touitou E. Scleroglucan sustained-release oral preparations. Part II. Effects of additives. Drug Des Deliv 1990; 5: 249-57.

[68]   Touitou E, Alhaique F, Riccieri FM, Riccioni G, Santucci E. Scleroglucan sustained-release oral preparations. Part I. *In vitro* experiments. Drug Des Deliv 1989; 5: 141-8.

[69]   Francois NJ, Daraio ME. Preparation and characterization of scleroglucan drug delivery films: the effect of freeze-thaw cycling. J Appl Polym Sci 2009; 112: 1994-2000.

[70]   Alhaique F, Riccieri FM, Santucci E, Carafa M. Environmental effects on the delivery of drugs from a pH-sensitive matrix. Acta Technol Legis Med 1990; 1: 1-9.

[71]   Alhaique F, Riccieri FM, Santucci E, Crescenzi V, Gamini A. A possible pH-controlled drug-delivery system based on a derivative of the polysaccharide scleroglucan. J Pharm Pharmacol 1985; 37: 310-13.

[72]   Casadei MA, Matricardi P, Fabrizi G, Feeney M, Paolicelli P. Physical gels of a carboxymethyl derivative of scleroglucan: Synthesis and characterization. Eur J Pharm Biopharm 2007; 67: 682-689.

[73]   Coviello T, Alhaique F, Parisi C, Matricardi P, Bocchinfuso G, Grassi M. A new polysaccharidic gel matrix for drug delivery: preparation and mechanical properties. J Controlled Release 2005; 102: 643-656.

[74]   Corrente F, Matricardi P, Paolicelli P, Tita B, Vitali F, Casadei MA. Physical carboxymethylscleroglucan /calcium ion hydrogels as modified drug delivery systems in topical formulations. Molecules 2009; 14: 2684-2698.

[75]   Feeney M, Casadei MA, Matricardi P. Carboxymethyl derivative of scleroglucan: a novel thermosensitive hydrogel forming polysaccharide for drug delivery applications. J Mater Sci Mater Med 2009; 20: 1081-1087.

[76]   Alhaique F, Casadei MA, Coviello T, Moracci FM. Chemical gels of scleroglucan for controlled-release formulations. Acta Technol Legis Med 2001; 12: 185-190.

[77]   Coviello T, Grassi M, Rambone G, Alhaique F. A crosslinked system from Scleroglucan derivative: preparation and characterization. Biomaterials 2001; 22: 1899-1909.

[78] Alhaique F, Coviello T, Grassi M, *et al.* New hydrogels from scleroglucan. Proc Int Symp Controlled Release Bioact Mater 1999; 26: 915-916.

[79] Coviello T, Grassi M, Rambone G, *et al.* Novel hydrogel system from scleroglucan: synthesis and characterization. J Controlled Release 1999; 60: 367-378.

[80] Maeda H, Coviello T, Yuguchi Y, *et al.* Structural characteristics of oxidized scleroglucan and its network. Polym Gels Networks 1999; 6: 355-366.

[81] Casadei MA, Pitarresi G, Benvenuti F, Giannuzzo M. Chemical gels of scleroglucan obtained by cross-linking with 1,o-dicarboxylic acids: synthesis and characterization. J Drug Delivery Sci Technol 2005; 15: 145-150.

[82] Alhaique F, Riccieri FM, Santucci E, Crescenzi V. Oxidized scleroglucan for the design of a pH-controlled drug delivery system. Chim Oggi 1986; 7-8: 11-15.

[83] Marchetti F, Bergamin M, Bosi S, Khan R, Murano E, Norbedo S. Synthesis of 6-deoxy-6-chloro and 6-deoxy-6-bromo derivatives of scleroglucan as intermediates for conjugation with methotrexate and other carboxylate containing compounds. Carbohydr Polym 2009; 75: 670-676.

[84] Coviello T, Dentini M, Rambone G, *et al.* A novel co-crosslinked polysaccharide: studies for a controlled delivery matrix. J Cont Rel 1998; 55: 57-66.

[85] Alhaique F, Coviello T, Rambone G, *et al.* A gellan-scleroglucan co-crosslinked hydrogel for controlled drug delivery. Proc Int Symp Controlled Release Bioact Mater 1998; 25: 866-867.

[86] Lee C-M, Jeong H-J, Kim D-W, Lee K-Y. Alginate/carboxymethyl scleroglucan hydrogels for controlled release of protein drugs. Macromol Res 2008; 16: 429-433.

[87] Coviello T, Grassi M, Lapasin R, Marino A, Alhaique F. Scleroglucan/borax: characterization of a novel hydrogel system suitable for drug delivery. Biomaterials 2003; 24: 2789-2798.

[88] Coviello T, Coluzzi G, Palleschi A, Grassi M, Santucci E, Alhaique F. Structural and rheological characterization of Scleroglucan/borax hydrogel for drug delivery. Int J Biol Macromol 2003; 32: 83-92.

[89] Palleschi A, Coviello T, Bocchinfuso G, Alhaique F. Investigation on a new scleroglucan/borax hydrogel: Structure and drug release. Int J Pharm 2006; 322: 13-21.

[90] Coviello T, Grassi M, Palleschi A, *et al.* A new scleroglucan/borax hydrogel: swelling and drug release studies. Int J Pharm 2005; 289: 97-107.

[91] Colinet I, Picton L, Muller G, Le Cerf D. pH-dependent stability of scleroglucan borate gels. Carbohydr Polym 2007; 69: 65-71.

[92] Coviello T, Matricardi P, Balena A, Chiapperino B, Alhaique F. Hydrogels from scleroglucan and ionic crosslinkers: Characterization and drug delivery. J Appl Polymer Sci 2010; 115: 3610-3622.

[93] Matricardi P, Onorati I, Masci G, Coviello T, Alhaique F. Semi-IPN hydrogel based on scleroglucan and alginate: drug delivery behavior and mechanical characterisation. J Drug Delivery Sci Technol 2007; 17:193-197.

[94] Matricardi P, Onorati I, Coviello T, Alhaique F. Drug delivery matrices based on scleroglucan/alginate/borax gels. Int J Pharm 2006; 316: 21-28.

[95] Lee C-M, Lee H-C, Lee K-Y. O-Palmitoylcurdlan sulfate (OPCurS)-coated liposomes for oral drug delivery. J Biosci Bioeng 2005; 100: 255-259.

[96] Kim B-D, Na K, Choi H-K. Preparation and characterization of solid lipid nanoparticles (SLN) made of cacao butter and curdlan. Eur J Pharm Sci 2005; 24: 199-205.

[97] Subedi Robhash K, Kang Keon W, Choi H-K. Preparation and characterization of solid lipid nanoparticles loaded with doxorubicin. Eur J Pharm Sci 2009; 37: 508-13.

[98] Carafa M, Marianecci C, Annibaldi V, Di Stefano A, Sozio P, Santucci E. Novel O-palmitoylscleroglucan-coated liposomes as drug carriers: Development, characterization and interaction with leuprolide. Int J Pharm 2006; 325: 155-162.

[99] Mochizuki S, Sakurai K. A novel polysaccharide/polynucleotide complex and its application to bio-functional DNA delivery system. Polymer J 2009; 41: 343-353.

[100] Minari J, Kelly AM, Mochizuki S, Shinkai S, Sakurai K. Polysaccharide/DNA supermolecule: triple helix consisting of one polynucleotide and two polysaccharide chains and its application to oligo-DNA delivery interface: polynucleotide, polysaccharide, molecular recognition. Curr Trends Polymer Sci 2008; 12: 75-102.

[101] Shinkai S, Tamesue S, Hirose R, Sakurai K, Nagasaki T, Numata M, β(1-3)-Glucan/carborane complexes for neutron capture therapy and their preparation. JP 2008222585. 2008 Sep.

[102] Numata M, Tamesue S, Nagasaki T, Sakurai K, Shinkai S. β(1-3)-glucan schizophyllan can act as a one-dimensional host to arrange icosahedral carboranes. Chem Lett 2007; 36: 668-669.

[103] Sakurai K, Shinkai S. Molecular Recognition of Adenine, Cytosine, and Uracil in a Single-Stranded RNA by a Natural Polysaccharide: Schizophyllan. J Am Chem Soc 2000; 122: 4520-4521.

[104] Kimura T, Koumoto K, Mizu M, Kobayashi H, Sakurai K, Shinkai S. A novel technique to specifically separate RNAs by Schizopyllan. Nucleic Acids Res 2001; 1: 283-4.

[105] Sakurai K, Kimura T, Koumoto K, Kobayashi R, Shinkai S. Application of schizophyllan as a novel gene carrier. Nucleic Acids Res 2001; 1: 223-4.

[106] Sakurai K, Shinkai S. Novel DNA-polysaccharide triple helixes and their application to a gene carrier. JInclusion Phen Macrocyclic Chem 2001; 41: 173-178.

[107] Pack DW, Hoffman AS, Pun S, Stayton PS. Design and development of polymers for gene delivery. Nature Reviews Drug Discovery 2005; 4: 581-593.

[108] Kimura T, Koumoto K, Sakurai K, Shinkai S. Polysaccharides-polynucleotide complexes(III): a novel interaction between the β(1-3)-glucan family and the single-stranded RNA poly(C). Chem Lett 2000; 11: 1242-1243.

[109] Sakurai K, Mizu M, Shinkai S. Polysaccharide-Polynucleotide Complexes. 2. Complementary Polynucleotide Mimic Behavior of the Natural Polysaccharide Schizophyllan in the Macromolecular Complex with Single-Stranded RNA and DNA. Biomacromolecules 2001; 2: 641-650.

[110] Mizu M, Kimura T, Koumoto K, Shinkai S, Sakurai K. Thermally induced conformational-transition of polydeoxy adenosine in the complex with schizophyllan and the base-length dependence of its stability. Chem Commun 2001; 5: 429-430.

[111] Sakurai K, Iguchi R, Koumoto K, *et al.* Polysaccharide-polynucleotide complexes VIII. Cation-induced complex formation between polyuridylic acid and schizophyllan. Biopolymers 2002; 65: 1-9.

[112] Karinaga R, Mizu M, Koumoto K, *et al.* First observation by fluorescence polarization of complexation between mRNA and the natural polysaccharide schizophyllan. Chemistry & Biodiversity 2004; 1: 634-639.

[113] Mizu M, Koumoto K, Anada T, Sakurai K, Shinkai S. Antisense oligonucleotides bound in the polysaccharide complex and the enhanced antisense effect due to the low hydrolysis. Biomaterials 2004; 25: 3117-3123.

[114] Mizu M, Koumoto K, Anada T, *et al.* Enhancement of the antisense effect of polysaccharide- polynucleotide complexes by preventing the antisense oligonucleotide from binding to proteins in the culture medium. B Chem Soc Jpn 2004; 77: 1101-1110.

[115] Koumoto K, Kimura T, Sakurai K, Shinkai S. Polysaccharide-Poly nucleotide Complexes (IV): Anti-hydrolysis Effect of the Schizophyllan/ Poly(C) Complex and the Complex Dissociation Induced by Amines. Bioorganic Chem 2001; 29:178-185.

[116] Koumoto K, Kimura T, Mizu M, Sakurai K, Shinkai S. Polysaccharide polynucleotide complexes. 9. Chemical modification of schizophyllan by introduction of a cationic charge into the side chain which enhances the thermal stability of schizophyllan-poly(C) complexes. Chem Commun 2001; 19: 1962-3.

[117] Koumoto K, Kimura T, Mizu M, Kunitabe T, Sakurai K, Shinkai S. Polysaccharide-polynucleotide complexes. Part 12. Enhanced affinity for various polynucleotide chains by site-specific chemical modification of schizophyllan. J Chem Soc, Perkin Transactions 1 2002; 22: 2477-2484.

[118] Nagasaki T, Hojo M, Uno A, *et al.* Long-term expression with a cationic polymer derived from a natural polysaccharide: schizophyllan. Bioconjugate Chem 2004; 15: 249-259.

[119] Karinaga R, Koumoto K, Mizu M, Anada T, Shinkai S, Sakurai K. PEG-appended β(1-3)-D-glucan schizophyllan to deliver antisense-oligonucleotides with avoiding lysosomal degradation. Biomaterials 2005; 26: 4866-4873.

[120] Matsumoto T, Numata M, Anada T, *et al.* Chemically modified polysaccharide schizophyllan for antisense oligonucleotides delivery to enhance the cellular uptake efficiency. Biochim Biophys Acta, General Subjects 2004; 1670: 91-104.

[121] Mizu M, Koumoto K, Anada T, *et al.* A polysaccharide carrier for immunostimulatory CpG DNA to enhance cytokine secretion. J Am Chem Soc 2004; 126: 8372-8373.

[122] Sakurai K, Shinkai S. A novel polysaccharide DNA-carrier to delivery CpG motifs to endosome. Polymer Preprints 2004; 45: 457-458.

[123] Koumoto K, Mizu M, Anada T, Nagasaki T, Shinkai S, Sakurai K. Cholesterol-appended β-(1→3)-D-glucan schizophyllan for antisense oligonucleotides delivery to enhance the cellular uptake. B Chem Soc Jpn 2005; 78: 1821-1830.

[124] Hasegawa T, M, Matsumoto T, Numata M, *et al.* Lactose-appended schizophyllan is a potential candidate as a hepatocyte-targeted antisense carrier. Chem Commun 2004; 4: 382-383.

[125] Karinaga R, Anada T, Minari J, *et al.* Galactose-PEG dual conjugation of β-(1→3)-D-glucan schizophyllan for antisense oligonucleotides delivery to enhance the cellular uptake. Biomaterials 2006; 27: 1626-1635.

[126] Hasegawa T, Fujisawa T, Haraguchi S, *et al.* Schizophyllan-folate conjugate as a new non-cytotoxic and cancer-targeted antisense carrier. Bioorganic & Medicinal Chem Lett 2005; 15: 327-330.

[127] Anada T, Karinaga R, Mizu M, *et al.* Transferrin-appended β-(1→3)-D-glucan schizophyllan for antisense oligonucleotide delivery to enhance the cellular uptake. e-Journal of Surface Science and Nanotechnology 2005; 3: 195-202.

[128] Minari J, Kubo T, Ohba H, *et al.* Delivery of antisense oligonucleotides to nuclear telomere RNA by use of a complex between polysaccharide and polynucleotide. B Chem Soc Jpn 2007; 80: 1091-1098.

[129] Anada T, Karinaga R, Koumoto K, *et al.* Linear double-stranded DNA that mimics an infective tail of virus genome to enhance transfection. Journal of Controlled Release 2005; 108: 529-539.

[130] Takeda Y, Shimada N, Kaneko K, Shinkai S, Sakurai K. Ternary Complex Consisting of DNA, Polycation, and a Natural Polysaccharide of Schizophyllan to Induce Cellular Uptake by Antigen Presenting Cells. Biomacromolecules 2007; 8: 1178-1186.

[131] Shimada N, Ishii KJ, Takeda Y, *et al.* Synthesis and *in vitro* Characterization of Antigen-Conjugated Polysaccharide as a CpG DNA Carrier. Bioconjugate Chem 2006; 17: 1136-1140.

[132] Shimada N, Coban C, Takeda Y, *et al.* A Polysaccharide Carrier to Effectively Deliver Native Phosphodiester CpG DNA to Antigen-Presenting Cells. Bioconjugate Chem 2007; 18: 1280-1286.

[133] Frey A, Frey A, Giannasca KT, *et al.* Role of the glycocalyx in regulating access of microparticles to apical plasma membranes of intestinal epithelial cells: implications for microbial attachment and oral vaccine targeting. J Exp Med 1996; 184: 1045-1059.

[134] Aouadi M, Tesz GJ, Nicoloro SM, *et al.* Orally delivered siRNA targeting macrophage Map4k4 suppresses systemic inflammation. Nature 2009; 458: 1180-1184.

[135] Young S-H, Ostroff GR, Zeidler-Erdely PC, Roberts JR, Antonini JM, Castranova V. A Comparison of the Pulmonary Inflammatory Potential of Different Components of Yeast Cell Wall. J Toxicol Environ Health, Part A 2007; 70: 1116-1124.

[136] Jamas S, Ostroff GR, Easson DD, Jr., Glucan drug delivery system and adjuvant. WO 9015596. 1990 Dec.

[137] Ostroff GR, Easson DD, Jr., Jamas S. A new β-glucan-based macrophage-targeted adjuvant. ACS Symp Ser 1991; 469: 52-9.

[138] Ostroff GR, Easson DD, Jr., Jamas S. Macrophage-targeted polysaccharide microcapsules for antigen and drug delivery. Polymer Preprints 1990;31(2):200-201.

[139] Ostroff GR, Drug delivery product and methods. US 2005281781. 2005 Dec.

[140] Czech MP, Ostroff GR. Nanoparticles encapsulated within yeast cell wall microparticles for nucleic acid delivery across the intestinal wall to macrophages and dendritic cells for transient gene therapy. WO 2009058913. 2009 May.

[141] Soto E, Ostroff GR. Characterization of Multilayered Nanoparticles Inside Yeast Cell Wall Particles for DNA Delivery. Bioconjugate Chem 2008; 19: 840-848.

[142] Soto E, Ostroff G. Oral macrophage mediated gene delivery system. NSTI Nanotech Technical Proceedings 2007; 2: 378-381.

[143] Soto E, Ostroff GR. Glucan Particles as an Efficient siRNA Delivery Vehicle. NSTI Nanotech Technical Proceedings 2008; 2: 332-335.

[144] Soto ER, Ostroff GR. Yeast cell wall particles as a versatile macromolecular delivery system. PMSE Prepr. 2008; 98: 591-593.

[145] Ostroff GR, Tipper DJ. Yeast cell particles with reduced mannan in cell wall for use as oral delivery vehicles for antigen or vaccine. WO 2007109564. 2007 Sep.

[146] Franklin L, Ostroff G. Compositions containing a hollow glucan particle or a cell wall particle encapsulating a terpene component for treatment of infections in plants and animals. WO 2005113128. 2005 May.

[147] Franklin L, Ostroff G. Terpene nematocides. WO 2005070213. 2005 May.

[148] Franklin L, Ostroff G, Harman G, Terpene mixtures as bactericides and fungicides for agriculture and medicine. WO 2007063268. 2007 Jun.

[149] Franklin L, Ostroff G, Harman G. Pesticidal terpenes encapsulated into hollow glucan particles. WO 2007063267. 2007 Jun.

[150] Plank C, Zatloukal K, Cotten M, Mechtler K, Wagner E. Gene Transfer into Hepatocytes Using Asialoglycoprotein Receptor Mediated Endocytosis of DNA Complexed with an Artificial Tetra-Antennary Galactose Ligand. Bioconjugate Chem 1992; 3: 533-539.

CHAPTER 6

Detrimental Effects of β(1-3),(1-6)-D-Glucans

Naohito Ohno*

Laboratory for Immunopharmacology of Microbial Products, School of Pharmacy, Tokyo University of Pharmacy & Life Science, 1432-1 Horinouchi, Hachioji, Tokyo 192-0392, Japan

Abstract: Beta-glucans (BG) are major cell wall components of fungi and are also found in plants and some bacteria. BG are a heterogeneous group of polymers, consisting of β(1-3) linked and β(1-6) linked β-D-glucopyranosyl units. As they are not found in animals, BG are considered to have pathogen-associated molecular patterns (PAMPs) and are recognized by the innate and acquired immune systems. BG show various activities, both beneficial and detrimental to the host. This chapter focuses on the detrimental effects of BG. It was found that BG induced inflammation, sepsis, and rheumatoid arthritis. BG also have the characteristics of an antigen, strongly supporting BG as representative PAMPs from fungi.

INTRODUCTION

Beta-glucans (BG) are major cell wall components of fungi and are also found in plants and some bacteria. BG are a heterogeneous group of polymers, consisting of β(1-3) linked and β(1-6) linked β-D-glucopyranosyl units. Both of the linkages form various segments to build up macromolecules having complex architecture. Among BG, the 6-branched 1,3-BG is the best characterized. As they are not found in animals, these glucans are considered to have pathogen-associated molecular patterns (PAMPs) [1] and are recognized by the innate and acquired immune systems. Some BG are applied clinically and are well-known biological response modifiers. We and others have demonstrated that the immunomodulating activity of BG is related to their effects on both humoral and cellular immunity. *i.e.*, activation of complement produced anaphylatoxin peptides, and activation of immune effecter cells, such as macrophages, neutrophils, and dendritic cells induced various cytokines, chemokines, and growth factors [2-4].

Host defense against microbial attack and against spontaneously arising malignant tumor cells involves a dynamic orchestrated interplay of innate and acquired immune responses, and PAMPs can initiate these systems. It is still unclear how BG mediate their effects. Recent studies, however, are starting to shed some light on the cellular receptors on BG and their molecular mechanisms. These receptors function to recognize and eliminate pathogenic fungi, such as *Aspergillus fumigatus*, *Candida albicans*, and *Pneumocystis jirovecii*, which generally contain BG in their cell walls [5].

In contrast, BG are released into the blood during infection. A soluble form of BG is known to show various biological activities, which might also be strongly related to the pathogenesis of fungal infection. The host defense system of limulus amoebocytes contains coagulation systems against both bacterial endotoxins and fungal BG. In addition, plants as well as insects have various BG-binding proteins as a component of their host defense systems. From these findings, it is of note that BG have both beneficial and detrimental effects. Thus, there is no doubt that BG have the characteristics of a "toxin". There is evidence supporting this concept as follows: Zymosan is a particle of baker's yeast origin and a well-known reagent for research on the acute inflammatory response; BG caused hyperinflammation and necrosis in mice carrying phagocyte NADPH oxidase deficiency; BG were deposited in the reticuloendothelial organs for life; BG acted as an inducer of autoimmune diseases, and so on. Thus, it is worth analyzing BG from the viewpoint of a toxin, hazard, and harmful material. This chapter focuses on the detrimental characteristics of BG.

DETRIMENTAL EFFECT OF BG FROM THE VIEWPOINT OF ZYMOSAN

Zymosan, a crude BG preparation, is a well-known reagent inducing strong inflammatory responses [6]. Zymosan is prepared from yeast, *Saccharomyces cerevisiae*, as a particle prepared from the debris of traditional extraction

---

***Address correspondence to: Dr. Naohito Ohno:** Laboratory for Immunopharmacology of Microbial Products, School of Pharmacy, Tokyo University of Pharmacy & Life Science, 1432-1 Horinouchi, Hachioji, Tokyo 192-0392, Japan E-mail: ohnonao@ps.toyaku.ac.jp

Vaclav Vetvicka and Miroslav Novak (Eds)

technologies with aqueous solution. Zymosan is composed of a mixture of cellular components, such as proteins, lipids, nucleic acids, and cell wall components. BG and mannan are the major cell wall components of yeast, act as PAMPs, and have been implicated in the recognition of yeast by the innate immune system. The mannan part is localized at the outer part of the cell, is highly soluble in water, and mainly released during extraction. Thus, the major cell wall components of zymosan are BG. Zymosan shows various biological activities related to inflammatory responses, and more than 7,000 papers can be found in PubMed. Zymosan shows various inflammatory reactions, both *in vivo* and *in vitro*. During such inflammation, various mediators are induced, such as lipid mediators, reactive oxygen species, cytokines, chemokines, apoptosis inducers, proteases, protease inhibitors, neuropeptides, and so on. Some of the activities, such as the synthesis of reactive oxygen, and inflammatory cytokines, are related to the phagocytic response of macrophages and neutrophils to microbes. The most striking feature of zymosan *in vivo* is its contribution to establishing animal disease models, such as "zymosan-induced multiple organ dysfunction syndrome", "zymosan-induced septic shock-like syndrome", "zymosan-induced generalized inflammation", "zymosan-induced peritonitis", "zymosan-induced rat model of acute bowel inflammation", "zymosan-induced earlobe inflammation in mice", and so on [7-14]. During analyses of these disease models, various important evidence for medical as well as biochemical research has been identified. Thus, zymosan is utilized as a model substance for analyzing acute inflammatory responses and the development of anti-inflammatory substances. As mentioned above, the major cell wall component of zymosan is BG; thus, many BG research groups have used zymosan as a BG preparation even though it is crude. At least in part, research has succeeded in identifying zymosan as an agonist of the functional BG receptor, and thus zymosan is the best example for discussing the side effects, or detrimental effects of BG.

To understand the molecular mechanism of BG activity, receptor molecules have been extensively searched, and currently three receptor molecules have been identified, the complement receptor (CR) 3, dectin-1, and lactosylceramide [15-20]. Recently, dectin-1-deficient mice have been established and applied in various studies [21]. Saijo *et al.* demonstrated that macrophages from dectin-1-deficient mice equally respond to zymosan to induce proinflammatory cytokines comparable to wild-type macrophages, whereas macrophages from MyD88-/- mice had decreased activity in cytokine production in response to zymosan. Furthermore, purified BG particles, OX-Zym, derived from zymosan showed no response in dectin-1-/mouse-derived macrophages and bone marrow-derived dendritic cells. These findings strongly suggested that the parent zymosan is a particle bearing multiple ligands for the receptors involved in macrophage activation. It was reported that both Toll-like receptor (TLR)-2 and TLR-6 are required for the activation of nuclear factor (NF)-κB and the production of inflammatory cytokines, such as tumor necrosis factor (TNF)-α by zymosan, and that zymosan enhances TLR-2-mediated NF-κB activation on coexpression of TLR-2 and dectin-1 [22]. Additionally, the binding of zymosan to macrophages was reported to be inhibited by mannan through the mannose receptor DC-SIGN [2]. Inoue *et al.* demonstrated that bronchial administration or inhalation of OX-BG induced lung inflammation [23-25]. Straszek *et al.* demonstrated the acute effect of BG-spiked office dust on nasal and pulmonary inflammation in guinea pigs [26]. Zymosan is a well-known activator of complement, especially through the alternative complement pathway. The resulting anaphylatoxins are also well known as inflammatory mediators. The above findings strongly suggest that BG have notable detrimental characteristics.

MOLECULAR MECHANISMS OF BG-INDUCED INFLAMMATORY RESPONSES

Cytokines and their networks are the key to immune as well as inflammatory responses. This section focuses on BG-induced cytokine production in DBA/2 mice. The BG used in this section was SCG prepared from a fungus, *Sparassis crispa* [27]. From the series of studies, SCG was found to produce various cytokines, such as IL-6, IL-12, IFN-γ, TNF-α, and GM-CSF, in an *in vitro* spleen cell culture. In the process of analyzing the mechanism, it was apparent that GM-CSF is a key factor in reactivity to BG, and GM-CSF induction by SCG is a specific step for other cytokine inductions *in vitro*. In fact, neutralizing GM-CSF in the splenocyte culture significantly inhibited other cytokine production, and a similar conclusion was reached using recombinant GM-CSF [28]. These results supported the concept that GM-CSF is essential in splenocytes for reactivity to BG (Fig. **1**).

GM-CSF is a 23-kD glycoprotein known as a hematopoietic growth factor required for the proliferation and survival of hematopoietic cells committed to granulocytic and macrophage cell lineages and myeloid leukemic cells [29]. It is also required for the differentiation of these cells into neutrophilic or eosinophilic granulocytes, macrophages, bone marrow macrophages, or dendritic cells [30-32]. In addition to these physiological roles, a growing body of evidence indicates that GM-CSF has important functions in host response to external stimuli and inflammatory/autoimmune conditions

[33]. Numerous studies in rodents have indicated the positive action of GM-CSF against many types of bacteria and fungi [34]. GM-CSF-deficient mice develop normally and show no major perturbation of hematopoiesis. On the other hand, studies using GM-CSF-deficient mice showed that GM-CSF action in the steady state was important for alveolar macrophage maturation in mice [35]. BG appear to be responsible for at least part of the stimulus in fungal infection in the lung [36]. These results imply a relationship between alveolar macrophages and BG in fungal infection in the lung. Willment *et al.* reported that BG receptor : dectin-1 expression and function are enhanced in GM-CSF-treated macrophages [37]. These findings suggested that the level of expression of BG receptor on leukocytes regulated by GM-CSF would modulate the responsiveness to BG.

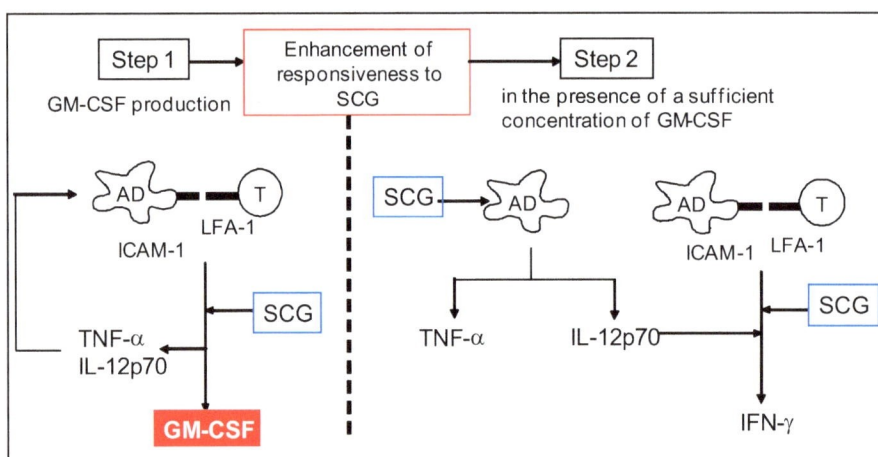

Figure 1: Model of the mechanism of cytokine induction by SCG in DBA/2 mice

To further analyze the molecular mechanisms of BG-induced inflammatory responses, DNA-microarray analysis was performed using the spleen cell culture of DBA/2 mice stimulated with SCG. Hundreds of genes have been specifically induced by SCG, and some of the significantly induced genes are summarized in Table **1**. Genes of adhesion molecules, chemokines, cytokines, hormones, receptors, as well as various enzymes have been induced. These findings strongly suggest that BG-induced immune modulation is not a simple/single reaction cascade but multiple/various signaling events. Further analyses might be necessary for precise understanding.

Table 1: Genes induced by SCG in spleen cell culture of DBA/2 mice

==

intercellular adhesion molecule 1 (ICAM-1) mRNA,

CD14 antigen (Cd14),

CD9 antigen (Cd9),

integrin beta 3 (Itgb3),

CD69 antigen (Cd69),

chemokine (C-C motif) ligand 2 (Ccl2),

chemokine (C-C motif) ligand 25 (Ccl25),

chemokine (C-C motif) ligand 3 (Ccl3),

chemokine (C-C motif) ligand 4 (Ccl4),

chemokine (C-C motif) receptor-like 2 (Ccrl2),

chemokine (C-X-C motif) ligand 1 (Cxcl1),

chemokine (C-X-C motif) ligand 10 (Cxcl10),

chemokine (C-X-C motif) ligand 2 (Cxcl2),

interferon gamma (Ifng),

interferon-induced protein with tetratricopeptide repeats 1 (Ifit1),

interferon-induced protein with tetratricopeptide repeats 2 (Ifit2),

interferon-induced protein with tetratricopeptide repeats 3 (Ifit3),

interleukin 1 alpha (Il1a),

interleukin 1 beta (Il1b),

interleukin 1 receptor antagonist (Il1rn),

interleukin 22 (Il22),

interleukin 6 (Il6),

tumor necrosis factor (Tnf),

heparin-binding EGF-like growth factor (Hbegf),

coagulation factor III (F3),

endothelin 1 (Edn1),

ferredoxin 1 (Fdx1),

glutathione peroxidase 1 (Gpx1),

heat shock protein 1A (Hspa1a),

immune-responsive gene 1 (Irg1)

gastrin (Gast),

inhibin beta-A (Inhba),

prothymosin alpha (Ptma),

C-type lectin domain family 4, member d (Clec4d),

C-type lectin domain family 4, member e (Clec4e),

C-type lectin domain family 4, member n (Clec4n),

somatostatin receptor 2 (Sstr2),

adenylate cyclase 1 (Adcy1),

acyl-CoA synthetase long-chain family memb.4,

ADP-ribosylation factor-like 4 (Arl4),

polymerase (RNA) II (DNA directed) polypeptide H (Polr2h),

a disintegrin-like and metalloprotease (reprolysin type) with thrombospondin type 1 motif, 20 (Adamts20),

ATP synthase, H+ transporting, mitochondrial F0 complex, subunit c (subunit 9), isoform 3 (Atp5g3),

amine oxidase, flavin containing 1 (Aof1),

calcium/calmodulin-dependent serine protein kinase (MAGUK family) (Cask),

cyclin-dependent kinase 8 (Cdk8), transcript variant 2,

heme oxygenase (decycling) 1 (Hmox1),

hexokinase 2 (Hk2),

hypoxanthine guanine phosphoribosyl transferase 1 (Hprt1),

lactate dehydrogenase 1, A chain (Ldh1),

mitogen-activated protein kinase kinase kinase 8 (Map3k8),

ornithine decarboxylase, structural 1 (Odc1),

phosphoserine phosphatase (Psph),

polo-like kinase 2 (Drosophila) (Plk2),

prostaglandin-endoperoxide synthase 2 (Ptgs2),

protein phosphatase 1, catalytic subunit, beta isoform (Ppp1cb),

protein phosphatase 4, regulatory subunit 2 (Ppp4r2),

ubiquitin specific protease 1 (Usp1),

ubiquitin-conjugating enzyme E2B, RAD6 homology (S. cerevisiae) (Ube2b),

thioredoxin reductase 1 (Txnrd1),

thioredoxin-like 1 (Txnl1),

sterol O-acyltransferase 1,

protein kinase inhibitor (testicular isoform) mRNA,

autocrine motility factor receptor (Amfr),

inositol 1,4,5-triphosphate receptor 1 (Itpr1),

lamin B receptor (Lbr),

oxidized low density lipoprotein (lectin-like) receptor 1 (Olr1),

poliovirus receptor (Pvr),

protein C receptor, endothelial (Procr),

fos-like antigen 2 (Fosl2),

Jun-B oncogene (Junb),
mRNA for glucose transporter type 3.
synuclein, alpha (Snca),
kinesin family member C5A (Kifc5a),
matrin 3 (Matr3),

==

EFFECT OF BG ON AUTOIMMUNE DISEASES

Rheumatoid arthritis (RA) is one of the most common systemic autoimmune diseases. It is characterized by chronic joint inflammation, the formation of a rheumatoid pannus and eventually, tissue degradation and joint destruction [38]. Although new and effective therapeutic approaches are being developed continuously, the etiology of RA is still unknown. Animal models of RA are used to investigate the pathogenesis and to develop therapeutic drugs. For example, collagen-induced arthritis (CIA) is an animal model of RA [39,40]. CIA is induced by inoculation with type II collagen (CII) emulsified with Freund's complete adjuvant (FCA), followed by a booster injection [40]. In the classical CIA model, FCA is essential for the induction of arthritis in mice. Recently, we have reported that BG derived from *C. albicans* acted as an adjuvant for CIA [41]. We used a BG, named OX-CA, prepared from *C. albicans* by oxidation with sodium hypochlorite (NaClO). The results suggested that not only bacterial components in FCA but also BG derived from *C. albicans* acted as an adjuvant for the induction of CIA.

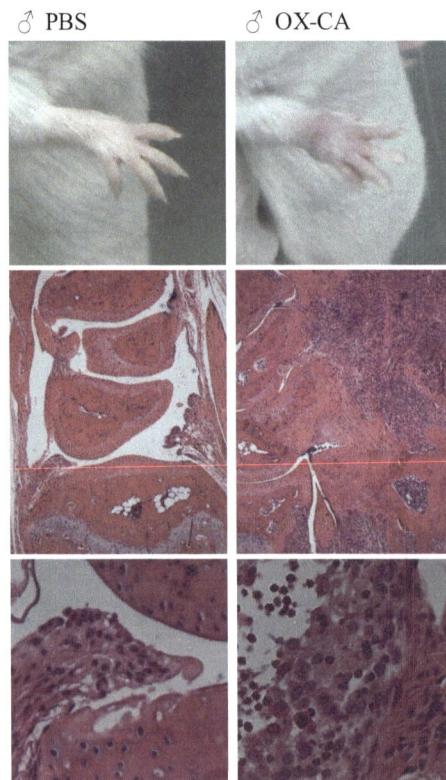

Figure 2: Severity of arthritis induced by injection of BG derived from *Candida albicans* in SKG mice. Non-arthritic joint of a male mouse injected with PBS (left). Joints of a SKG male mouse injected with OX-CA (right). HE staining of these joints, Middle, x 10, Bottom, x 40

The SKG strain is a autoimmune mouse developing chronic autoimmune arthritis. The symptoms of SKG, such as bilateral swelling of the finger, wrist and ankle joints, and the production of various autoantibodies, resemble human RA. SKG mice carry a mutation of the gene encoding an SH2 domain of ZAP-70, a key signal transduction molecule for T-cells [42]. SKG mice develop arthritis at about 2 months of age under conventional conditions; however, in a strictly controlled specific pathogen-free environment, SKG mice failed to develop chronic arthritis.

These results suggested that infection from environmental agents, especially fungal infections, is a trigger for the induction of arthritis in SKG. In fact, SKG mice injected with BG, such as laminarin and curdlan, developed arthritis under SPF conditions [43]. *C. albicans* is the most frequently isolated opportunistic fungal pathogen and is reported to have a role in the pathogenesis of arthritis [44-46]. Therefore, to assess whether BG derived from pathogenic fungi, such as *C. albicans* induces arthritis in SKG mice, we administered particulate BG (OX-CA) or soluble BG (CSBG) prepared from *C. albicans* [47]. SKG mice injected with OX-CA or CSBG developed more severe arthritis earlier than with LAM in both genders. Arthritis in female SKG injected with OXCA or CSBG was more serious than that in male mice, for example, not only were wrists joints swollen but also all fingers (Fig. **2**).

We previously studied the relationship between the structure and activity of various BG [48]. OX-CA and CSBG showed various biological activities, such as IL-6 synthesis of macrophages *in vitro*, anti-tumor effects, TNF-α production from RAW264.7 cells, and adjuvant effects for the induction of CIA, whereas LAM hardly showed those activities. The difference in biological activities induced with BG might be related to the development of arthritis in SKG. We also examined the ability of other BG preparations. Schizophyllan (SPG) from *Schizophyllum commune* is a fungal BG used clinically in Japan for cervical cancer patients in combination with irradiation to enhance the immunological surveillance system [49]. SPG also induced arthritis in SKG mice, and the arthritis was more severe and with higher incidence than with LAM.

INDUCTION OF SEPTIC SHOCK BY CONCOMITANT USE OF NSAIDS AND BG

Sepsis is an infection-induced syndrome resulting in a systemic inflammatory response that is complicated by the dysfunction of multiple organs [50,51]. During the course of studying the mechanisms of BG-mediated immune/inflammatory responses, we have found that concomitant administration of BG and non-steroidal anti-inflammatory drug, NSAIDs, induced the sudden death of mice [52]. This observation was fully unexpected and a possible BG side-effect, so we have analyzed further. We finally identified sepsis induced by damage to the mucosal membrane and the resulting transmigration of intestinal bacteria into the blood stream [53]. Symptoms of sepsis have been induced in this model, such as hypercytokinemia, hypoglycemia, liver dysfunction, detection of bacteria in various organs, accumulation of leukocytes in major organs, increased production of superoxide and nitrogen oxide, and so on. We also found that other BG preparations, and whole cells of fungi, yeasts, and bacteria induced similar sepsis [54].

We have analyzed this model from various points of view. The C3H/HeJ strain is a well characterized strain of mice showing significantly low sensitivity to LPS due to a point mutation in TLR-4. We have applied the SPG/NSAID model for C3H/HeJ mice and found that the magnitude of enhanced mortality in this strain was similar to ICR, a wild-type LPS-sensitive strain, although the timing was slightly delayed (Figs. **3** and **4**) [55]. This strongly suggested that the LPS and subsequent TLR-4-mediated inflammatory signaling might not be the sole molecule responsible for septic shock in this model.

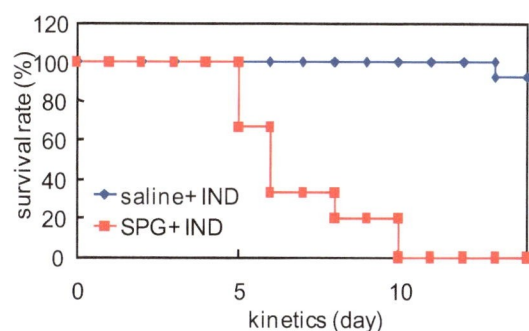

Figure 3: Survival of SPG/IND-administered C3H/HeJ mice

Thus, we tested other PAMPs for inflammatory cytokine synthesis *in vitro* and found that either Pam3 or CpG could directly induce IL-6 from spleen cell culture and PEC culture. This strongly suggested that these PAMPs stimulate inflammatory responses through other TLRs. In addition, we tested the cooperative/synergistic effects of other

PAMPs, Pam3 and CpG, to the cells of C3H/HeJ mice, and found synergistic effects producing inflammatory cytokines. This strongly suggested that endogenous septic shock induced by the microbial translocation of intestinal flora might also occur in C3H/HeJ mice with the resulting enhanced response to PAMPs stimulating TLRs, leading to death. Alternatively, the side-effect of BG might induce various microbial infections.

We have applied this model to analyze the therapeutic protocol using antibiotics, fradiomycin and polymyxin B, and demonstrated increased survival days, but the rate was not changed. Nameda *et al.* have studied the therapeutic protocol using ABPC/IPM/CS and LCM [56]. ABPC/IPM/CS are β-lactam derivative and broad-spectrum antibiotics. LCM is effective against Gram-positive and anaerobic bacteria, but is not effective against Gram-negative bacillus such as *Escherichia coli* [57]. It was found that the effect of these antibiotics on the survival rate was significantly different. The rate of survival was improved by LCM administration. LCM treatment did not decrease the body weight of SPG/IND-administered mice. Moreover, SPG/IND/LCM-treated mice did not exhibit intra-abdominal adhesion and intestinal injury, and the effect continued to the last stage. Nevertheless, ABPC/IPM/CS treatment showed strong bactericidal activity compared with LCM, and selective sterilization by LCM may lead to a successful increase of the survival rate.

Using this model, we have learned about the anti-inflammatory effect of macrolides. *i.e.*, production of reactive oxygen species by neutrophils and macrophages, secretion of pro-inflammatory cytokines, adhesion molecule expression on epithelial cells, and leukocyte adhesion and infiltration into tissues. The most probable mechanism of macrolides to induce anti-inflammatory properties is inhibition of two of the most important transcription factors, nuclear factor kappa B (NF-κB) and activator protein 1 (AP-1).

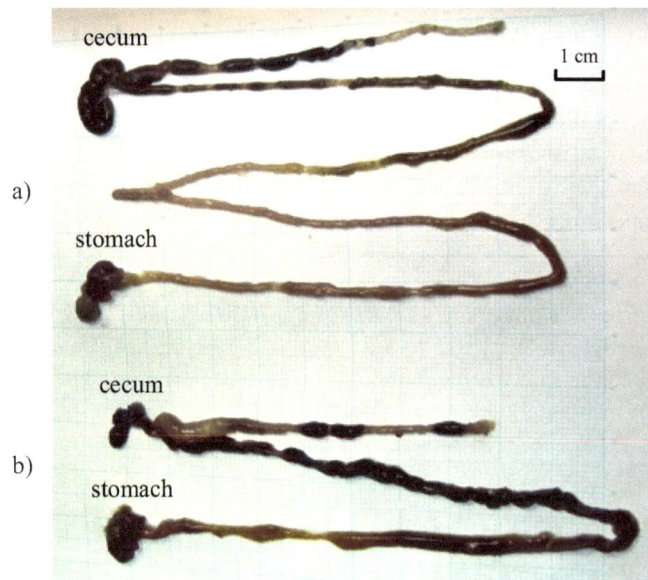

Figure 4: Macroscopic view of the intestines of SPG/IND-administered C3H/HeJ mice. A: saline/IND, b: SPG/IND

Clindamycin (CLDM), a synthetic analogue of LCM, is effective against Gram-positive and anaerobic bacteria, but is not effective against *E. coli*. However, it has been reported that CLDM protected mice from sepsis by inhibiting pro-inflammatory cytokine synthesis of leukocytes stimulated by LPS. CLDM/LCM is known to modulate inflammatory cytokine induction in LPS-stimulated mouse peritoneal macrophages. CLDM/LCM enhanced the cytoplasmic function of polymorphonuclear leukocytes, accelerated phagocytosis and promoted antibody production by lymphocytes. It is also reported that subminimal inhibitory concentrations of CLDM influenced bacterial viability, toxin production, and host response. These studies strongly supported that LCM may suppress the increase of LPS sensitivity in SPG/IND-induced septic shock. Additionally, PSL, a steroid used to treat severe infectious diseases and septic shock, increased the survival rate of SPG/IND-treated mice. These findings suggested that the toxicity induced by SPG/IND administration depends on the excessive response to endogenous cytokines in response to normal bacterial flora spreading in the host.

We have analyzed this model from the viewpoint of drug-metabolizing enzymes. The cytochrome P450 (CYP) superfamily is a family of enzymes that catalyze numerous reactions, including fatty acid metabolism, xenobiotic biotransformation, and endogenous substrate synthesis and metabolism [58]. It is known that the mediators involved in inflammation and infection can cause changes in the activities and expression of the CYP enzyme system. A large number of reports have shown that CYP and their activities are mostly suppressed in animal models of endotoxemia [59].

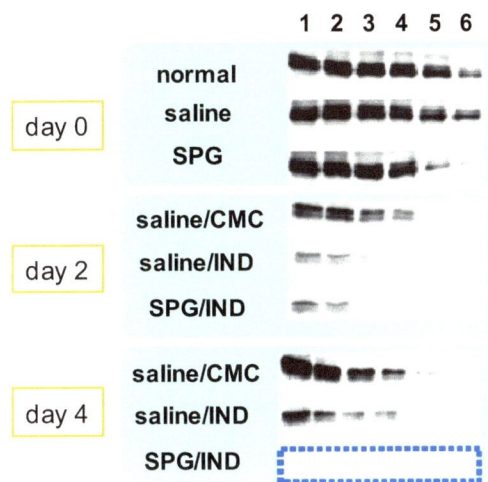

Figure 5: CYP3A11 expression on SPB/IND-administered C3H/HeJ mice. Protein conc. (µg/ml); lane 1 to 6: 500, 250, 125, 62.5, 31.25, 15.625.

The various CYP isoforms primarily involved in drug metabolism are grouped into CYP1, CYP2, and CYP3 subfamilies [60]. We examined the change in CYP3A11 and CYP2E1 expressions in this model. Both CYP3A11 and CYP2E1 activities decreased by LPS stimulation but had different kinetics of protein expression by inflammatory cytokine modulation [61]. In this study we have found that both CYP3A11 and CYP2E1 decreased but much more markedly in CYP3A11 (Fig. **5**). The sensitivity of CYP3A11 and CYP2E1 was also different between ICR and C3H/HeJ. These differences might be regulated by various genetic backgrounds directly and indirectly, such as the polymorphism of promoter and coding region sequences, and might merit further investigation. Eum *et al.* reported that CYP2E1 was decreased in the CLP model and this reduction was recovered by the administration of a nitric oxide inhibitor [62]. We have reported that nitric oxide production was increased in SPG/IND-treated ICR mice [53]. Nitric oxide plays a role as a protective factor in bacterial translocation; therefore, nitric oxide production in mice would induce the reduction of CYP2E1. In this model we have shown that mortality was strongly related to the concentration of IND and a higher concentration shortened the survival period. During the experimental period, IND might have been at least partly metabolized by CYP. Failure of CYP might increase the blood concentration of IND, and enhance mortality.

DEPOSITION AND METABOLISM OF BG

Because it is difficult to comprehend that BG are absorbed by the body under normal physiological conditions, there is little knowledge of their metabolism and accumulation. Its metabolism and accumulation should be taken into consideration when a large number of fungi grow inside the body of a deep mycosis patient, or when BG are administered as antitumor agents. Further, it should be added that it is not known to what extent various drugs or medical devices are contaminated by BG.

To examine accumulation in the body, the kinetics and deposition were investigated using the Limulus factor G test, and isotope-labeled BG [63,64]. As a result, it was found that BG injected into mice were accumulated mainly in the liver and spleen for an extremely long time with little decomposition. Here, "extremely long" indicates a period as long as six months or a year, which corresponds to the survival time of mice. The deposited BG were gradually degraded by oxidative metabolism using reactive oxygen species produced by phagocytes in organs. Administration of BG is known to augment the function of macrophages against invading microbes; thus, it is of no doubt that the degradation of BG mainly followed oxidative degradation, also due to the absence of BG hydrolase in animals. In human studies, the blood concentration of BG remains high for years in cancer patients administered anticancer BG preparations. These data strongly suggest that the deposition of BG in organs is a general characteristic.

In contrast, we have experienced the difficulty of BG absorption by the oral route. From the viewpoints of deposition and the slow degradation rate *in vivo*, such difficulty persists for life. However, from the pharmacological point of view, BG are the most widely used food in health care and self-medication. Many studies have confirmed that oral administration of BG is beneficial for health and for the prevention of various diseases. We have also demonstrated that oral administration of BG is effective against multiple solid tumors of the allogeneic as well as syngeneic systems, demonstrates metastasis inhibitory effects, increases the responsiveness of spleen cells to ConA and LPS, increases NK activity, exhibits peritoneal and alveolar macrophage activation, and enhances IgA production. Moreover, these effects were strongly associated with Peyer's patch function [65,66]. Although conclusions cannot be drawn, detailed analyses of the absorption, accumulation, metabolism, and excretion of orally administered BG should be performed.

Another important view is the contamination of BG in drugs, especially drugs for injection. We therefore measured the concentration of BG in drugs for injection. We tested about 100 samples and found that the majority of drugs contained <10 pg/mL BG; however, some monoammonium glycyrrhizinates, elcatonins, and sodium ozagrels contained BG at >10 pg/mL (Fig. **6**). Currently, there have been no reports of clinical problems, but quality control of these medicines will be important in the future.

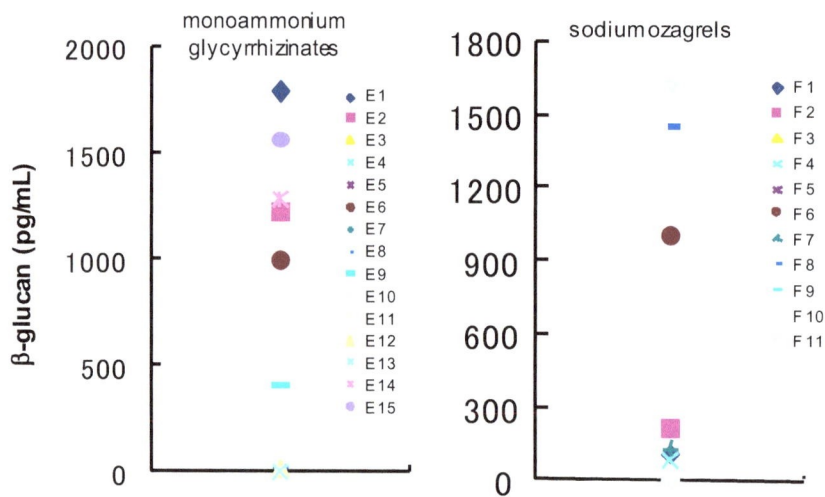

Figure 6: Beta-glucan concentration in drugs for injection. Concentration of beta-glucan was measured by Limulus factor G test. E 1-15 or F 1-11 represents concentration of each drug preparation.

ANTIBODY TO BG

BG have previously been categorized as low antigenic substances, and thus unlikely to induce antibody production. In practice, it is difficult to induce antibodies against anticancer BG, SPG. In addition, there have been many attempts to produce monoclonal antibodies against BG, but they have met with little success. In contrast, the anti-BG antibody was produced when mice were administered yeast cells [67]. These findings strongly suggested that SPG-type polysaccharides have difficulty in inducing antibodies, but not all BG have difficulty in inducing antibodies. For instance, when antibody titers against CSBG obtained from *Candida* were measured using human globulin preparations, IgG class antibody was detected with a considerable titer [68,69]. When anti-BG titers were measured in healthy volunteers, the antibodies were present in all volunteers. Moreover, when anti-BG titers were measured in cancer patients, autoimmune disease patients and mycosis patients, the titers were relatively decreased. Anti-BG antibodies were also observed in pigs, cows, horses and other animals. Of interest, antibodies against BGs were observed in mice, albeit only in specific strains [70]. These findings indicate that humans and animals are spontaneously sensitized to BG in food and the environment and react with acquired immunity as well as innate immunity. Activation of the immune system mediated by antibodies, complement and platelets is also important in discussing the immune as well as inflammatory mechanism of BG.

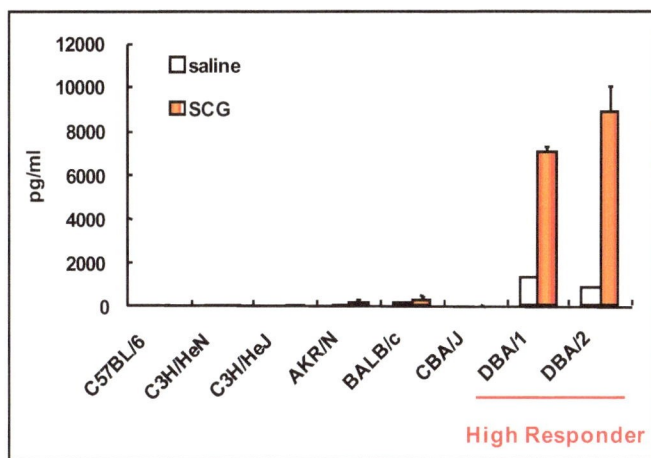

Figure 7: Strain difference of IFN-γ production by splenocytes stimulated with SCG in mice.

STRAIN AND INDIVIDUAL DIFFERENCES IN REACTIVITY TO BG

Animal studies, especially in the field of immune and inflammatory mechanisms, often use one or only a few inbred strains of mice. In some cases, gene knockout mice of the same background strains have been used. In contrast, it is well recognized that human genes contain a significant number of polymorphisms, thus resulting in a huge number of phenotypes. Studies on strain differences in the reactivity to BG are directly or indirectly important for the clarification of individual variations in humans, for both beneficial and detrimental effects. We have demonstrated that strain differences in the reactivity to BG, and DBA/1 and DBA/2 strains are highly sensitive to BG *in vitro* (Fig. 7) [71] and *in vivo* [72]. In an *in vitro* study, SCG induced high levels of IFN-γ, TNF-α, IL-12p70 and GM-CSF in splenocytes from DBA/1 and DBA/2 strains. Cytokine induction of SCG was not dependent on the gender or breeder. The level of IFN-γ induction by SCG gradually increased with the age of mice. In contrast, IFN-γ production was not induced by SCG *in vitro* in the splenocytes of almost all strains of mice, such as C57BL/6, BALB/c, AKR/N, CBA/J, C3H/HeN and C3H/HeJ. C57BL/6 x DBA/2 F1 hybrid mice exhibited little to no IFN-γ production [28]. These results indicate that DBA/1 and DBA/2 strains are highly sensitive to BG in a hereditary manner, conferred as a recessive genetic trait. This surprised us because generally the level of cytokine induction by pure soluble BG from leukocytes in naive mice is lower than that by other PAMPs.

The differences in genetic background between DBA/1 and DBA/2 mice are too large to be accounted for by mutation, and are probably due to substantial residual heterozygosity following crosses between substrains. DBA/1 and DBA/2 differ at least at the following loci: Car2, Ce2, Hc, H2, If1, Lsh, Tla, and Qa3. With such large differences, they should probably be regarded as different strains rather than substrains. DBA strains are widely used for immunological research *in vivo* and *in vitro*, *i.e.*, from the viewpoint of complement (C), DBA/1 mice are C5 normal, and DBA/2 mice are C5-deficient [73]. The MHC genotype restricts the antigen specificity of T cells, and MHC restriction is a key feature in the antigen recognition mechanism of T cells. Some diseases, such as collagen-induced arthritis, Crohn's disease and ulcerative colitis, have certain susceptibility genes in the major histocompatibility complex (MHC) region. The MHC haplotype of DBA/2 mice is H2d, as in BALB/c mice, while that of DBA/1 mice is H2q. These findings indicated that MHC restriction is not the major reason for the sensitivity to BG. Since large immunological differences are observed between DBA/1 and DBA/2 strains, it might be not easy to search for genes susceptible to the BG response. We believe that further analysis of homologous genes between DBA/1 and DBA/2 could clarify the genes controlling the susceptibility to BG.

As shown above, we thought that BG would be a weak immunogen; however, DBA/1 and DBA/2 mice also produced significantly higher titers of antibody to SCG than other inbred naive mice. On the other hand, the anti-SCG titer in each mouse differed significantly. The titer gradually increased with the age of mice. From these findings, in addition to the genetic background, environmental factors, such as food, air, and floor coverings, might influence these differences.

We have tested individual differences in BG responses in humans by measuring IL-8 and TNF-α synthesis of PBMC culture, and found more than 100-fold individual differences. In addition, the relative response of IL-8 and TNF-α production in each subject was significantly different. These results strongly suggested that the sensitivity of PBMC responses to BG varied significantly, similar to strain differences in mice. This again suggested that the beneficial and detrimental effects of BG might not be induced equally in individuals.

CONCLUSION

This chapter focused on the detrimental effects of BG. It was found that BG induced inflammation, sepsis, and rheumatoid arthritis. BG also have the characteristics of an antigen. This strongly supported BG as representative PAMPs from fungi. The detrimental effect of BG is marked, as shown above; however, from the clinical point of view, such effects currently do not attract attention. This is due to the lower incidence of fungal infection than bacterial and viral infections. From the viewpoint of structure, BG possess large heterogeneity, such as soluble vs. insoluble, branched vs. linear, single helix vs. triple helix, and the ratio of branching, as well as molecular weight. Such various differences might influence the host response. Further study is needed to fully understand the detrimental effects of BG.

REFERENCES

[1]   Janeway CA Jr, Medzhitov R. Innate immune recognition. Annu Rev Immunol 2002; 20: 197-216.
[2]   Taylor PR, Brown GD, Herre J *et al.* The role of SIGNR1 and the beta-glucan receptor (dectin-1) in the nonopsonic recognition of yeast by specific macrophages. J Immunol 2004; 172: 1157-62.
[3]   Herre J, Gordon S, Brown GD. Dectin-1 and its role in the recognition of beta-glucans by macrophages. Mol Immunol 2004; 40: 869-76.
[4]   Yadomae T. [Structure and biological activities of fungal β(1-3)-glucans] Yakugaku Zasshi 2000; 120: 413-31.
[5]   Medzhitov R, Janeway CA Jr. Decoding the patterns of self and nonself by the innate immune system. Science 2002; 296: 298-300.
[6]   Di Carlo FJ, Fiore JV. On the composition of zymosan. Science 1958; 127: 756-7.
[7]   Cash JL, White GE, Greaves DR. Chapter 17. Zymosan-induced peritonitis as a simple experimental system for the study of inflammation. Methods Enzymol 2009; 461: 379-96.
[8]   Dimitrova P, Gyurkovska V, Shalova I *et al.* Inhibition of zymosan-induced kidney dysfunction by tyrphostin AG-490. J Inflamm (Lond) 2009; 6: 13.
[9]   Au BT, Teixeira MM, Collins PD *et al.* Effect of PDE4 inhibitors on zymosan-induced IL-8 release from human neutrophils: synergism with prostanoids and salbutamol. Br J Pharmacol 1998; 123: 1260-6.
[10]   Au BT, Williams TJ, Collins PD. Zymosan-induced IL-8 release from human neutrophils involves activation via the CD11b/CD18 receptor and endogenous platelet-activating factor as an autocrine modulator. J Immunol 1994; 152: 5411-9.
[11]   Sanguedolce MV, Capo C, Bouhamdan M *et al.* Zymosan-induced tyrosine phosphorylations in human monocytes. Role of protein kinase C. J Immunol 1993; 151: 405-14.
[12]   Sanguedolce MV, Capo C, Bongrand P *et al.* Zymosan-stimulated tumor necrosis factor-alpha production by human monocytes. Down-modulation by phorbol ester. J Immunol 1992; 148: 2229-36.
[13]   Ridger VC, Pettipher ER, Bryant CE *et al.* Effect of the inducible nitric oxide synthase inhibitors aminoguanidine and L-N6 -(1-iminoethyl) lysine on zymosan-induced plasma extravasation in rat skin. J Immunol 1997; 159: 383-90.
[14]   Jiang X, Wu TH, Rubin RL. Bridging of neutrophils to target cells by opsonized zymosan enhances the cytotoxicity of neutrophil-produced H2O2. J Immunol 1997; 159: 2468-75.
[15]   Xia Y, Borland G, Huang J *et al.* Function of the lectin domain of Mac-1/complement receptor type 3 (CD11b/CD18) in regulating neutrophil adhesion. J Immunol 2002; 169: 6417-26.
[16]   Yan J, Vetvicka V, Xia Y *et al.* Beta-glucan, a "specific" biologic response modifier that uses antibodies to target tumors for cytotoxic recognition by leukocyte complement receptor type 3 (CD11b/CD18). J Immunol 1999; 163: 3045-52.
[17]   Ross GD, Vetvicka V, Yan J *et al.* Therapeutic intervention with complement and beta-glucan in cancer.

Immunopharmacology 1999; 42: 61-74.

[18]   Ozment-Skelton TR, deFluiter EA, Ha T *et al.* Leukocyte Dectin-1 expression is differentially regulated in fungal versus polymicrobial sepsis. Crit Care Med 2009; 37: 1038-45.

[19]   Palma AS, Feizi T, Zhang Y *et al.* Ligands for the beta-glucan receptor, Dectin-1, assigned using "designer" microarrays of oligosaccharide probes (neoglycolipids) generated from glucan polysaccharides. J Biol Chem 2006; 281: 5771-9.

[20]   Brown GD, Gordon S. Immune recognition. A new receptor for beta-glucans. Nature 2001; 413: 36-7.

[21]   Saijo S, Fujikado N, Furuta T *et al.* Dectin-1 is required for host defense against *Pneumocystis carinii* but not against *Candida albicans*. Nat Immunol 2007; 8: 39-46.

[22]   Gantner BN, Simmons RM, Canavera SJ *et al.* Collaborative induction of inflammatory responses by dectin-1 and Toll-like receptor 2. J Exp Med 2003; 197: 1107-17.

[23]   Inoue K, Takano H, Oda T *et al.* Candida soluble cell wall beta-D-glucan induces lung inflammation in mice. Int J Immunopathol Pharmacol 2007; 20: 499-508.

[24]   Inoue K, Takano H, Koike E *et al. Candida* soluble cell wall beta-glucan facilitates ovalbumin-induced allergic airway inflammation in mice: Possible role of antigen-presenting cells. Respir Res 2009; 10: 68.

[25]   Inoue K, Koike E, Yanagisawa R *et al.* Pulmonary exposure to soluble cell wall beta-(1, 3)-glucan of aspergillus induces proinflammatory response in mice. Int J Immunopathol Pharmacol 2009; 22: 287-97.

[26]   Straszek SP, Adamcakova-Dodd A, Metwali N *et al.* Acute effect of glucan-spiked office dust on nasal and pulmonary inflammation in guinea pigs. J Toxicol Environ Health A 2007; 70: 1923-8.

[27]   Ohno N, Miura NN, Nakajima M *et al.* Antitumor 1,3-beta-glucan from cultured fruit body of *Sparassis crispa*. Biol Pharm Bull 2000; 23: 866-72.

[28]   Harada T, Miura NN, Adachi Y *et al.* Granulocyte-macrophage colony-stimulating factor (GM-CSF) regulates cytokine induction by 1,3-beta-D-glucan SCG in DBA/2 mice in vitro. J Interferon Cytokine Res 2004; 24: 478-89.

[29]   Chen BD, Clark CR, Chou TH. Granulocyte/macrophage colony-stimulating factor stimulates monocyte and tissue macrophage proliferation and enhances their responsiveness to macrophage colony-stimulating factor. Blood 1988; 71: 997-1002.

[30]   Caux C, Dezutter-Dambuyant C, Schmitt D *et al.* GM-CSF and TNF-alpha cooperate in the generation of dendritic Langerhans cells. Nature 1992; 360: 258-61.

[31]   Inaba K, Inaba M, Romani N *et al.* Generation of large numbers of dendritic cells from mouse bone marrow cultures supplemented with granulocyte/macrophage colony- stimulating factor. J Exp Med 1992; 176: 1693-702.

[32]   Takahashi K, Miyakawa K, Wynn AA *et al.* Effects of granulocyte/macrophage colony- stimulating factor on the development and differentiation of CD5-positive macrophages and their potential derivation from a CD5-positive B-cell lineage in mice. Am J Pathol 1998; 152: 445-56.

[33]   Hamilton JA, Anderson GP. GM-CSF Biology. Growth Factors 2004; 22: 225-31.

[34]   Hubel K, Dale DC, Liles WC. Therapeutic use of cytokines to modulate phagocyte function for the treatment of infectious diseases: current status of granulocyte colony-stimulating factor, granulocyte-macrophage colony- stimulating factor, macrophage colony-stimulating factor, and interferon-gamma. J Infect Dis 2002; 185: 1490-501.

[35]   Stanley E, Lieschke GJ, Grail D *et al.* Granulocyte/macrophage colony-stimulating factor- deficient mice show no major perturbation of hematopoiesis but develop a characteristic pulmonary pathology. Proc Natl Acad Sci U S A 1994; 91: 5592-6.

[36]   Rapaka RR, Goetzman ES, Zheng M *et al.* Enhanced defense against *Pneumocystis carinii* mediated by a novel dectin-1 receptor Fc fusion protein. J Immunol 2007; 178: 3702-12.

[37]   Willment JA, Lin HH, Reid DM *et al.* Dectin-1 expression and function are enhanced on alternatively activated and GM-CSF-treated macrophages and are negatively regulated by IL-10, dexamethasone, and lipopolysaccharide. J Immunol 2003; 171: 4569-73.

[38]   Feldmann M, Brennan FM, Maini RN. Rheumatoid arthritis. Cell 1996; 85: 307-10.

[39]   Myers LK, Rosloniec EF, Cremer MA *et al.* Collagen-induced arthritis, an animal model of autoimmunity. Life Sci 1997; 61: 1861-78.

[40]   Courtenay JS, Dallman MJ, Dayan AD *et al.* Immunisation against heterologous type II collagen induces arthritis in mice. Nature 1980; 283: 666-8.

[41]   Hida S, Miura NN, Adachi Y *et al.* Effect of *Candida albicans* cell wall glucan as adjuvant for induction of autoimmune arthritis in mice. J Autoimmun 2005; 25: 93-101.

[42]   Sakaguchi S, Sakaguchi N, Yoshitomi H *et al.* Spontaneous development of autoimmune arthritis due to genetic anomaly of T cell signal transduction: Part 1. Semin Immunol 2006; 18: 199-206.

[43]   Yoshitomi H, Sakaguchi N, Kobayashi K *et al.* A role for fungal {beta}-glucans and their receptor Dectin-1 in the

induction of autoimmune arthritis in genetically susceptible mice. J Exp Med 2005; 201: 949-60.

[44] Yordanov M, Tchorbanov A, Ivanovska N. *Candida albicans* cell-wall fraction exacerbates collagen-induced arthritis in mice. Scand J Immunol 2005; 61: 301-8.

[45] Yordanov M, Danova S, Ivanovska N. Inflammation induced by inoculation of the joint with *Candida albicans*. Inflammation 2004; 28: 127-32.

[46] Dimitrova P, Yordanov M, Danova S *et al.* Enhanced resistance against systemic *Candida albicans* infection in mice treated with *C. albicans* DNA. FEMS Immunol Med Microbiol 2008; 53: 231-6.

[47] Hida S, Miura NN, Adachi Y *et al.* Cell wall beta-glucan derived from *Candida albicans* acts as a trigger for autoimmune arthritis in SKG mice. Biol Pharm Bull 2007; 30: 1589-92.

[48] Tokunaka K, Ohno N, Adachi Y *et al.* Immunopharmacological and immunotoxicological activities of a water-soluble (1-->3)-beta- D-glucan, CSBG from *Candida* spp. Int J Immunopharmacol 2000; 22: 383-94.

[49] Miyazaki K, Mizutani H, Katabuchi H *et al.* Activated (HLA-DR+) T-lymphocyte subsets in cervical carcinoma and effects of radiotherapy and immunotherapy with sizofiran on cell- mediated immunity and survival. Gynecol Oncol 1995; 56: 412-20.

[50] Friedman G, Silva E, Vincent JL. Has the mortality of septic shock changed with time. Crit Care Med 1998; 26: 2078-86.

[51] Dellinger RP, Schorr C. Severe sepsis in an emergency department: prevalence, rapid identification, and appropriate treatment. Crit Care Med 2007; 35: 2461-2.

[52] Takahashi H, Ohno N, Adachi Y *et al.* Association of immunological disorders in lethal side effect of NSAIDs on beta-glucan- administered mice. FEMS Immunol Med Microbiol 2001; 31: 1-14.

[53] Nameda Y, Miyoshi H, Tsuchiya K *et al.* Endotoxin-induced L-arginine pathway produces nitric oxide and modulates the Ca^{2+}-activated K^+ channel in cultured human dermal papilla cells. J Invest Dermatol 1996; 106: 342-5.

[54] Moriya K, Miura NN, Adachi Y *et al.* Systemic inflammatory response associated with augmentation and activation of leukocytes in Candida/indomethacin administered mice. Biol Pharm Bull 2002; 25: 816-22.

[55] Saito M, Nameda S, Miura NN *et al.* SPG/IND-induced septic shock in a LPS-low responder strain, C3H/HeJ mice. Microb Pathog 2008; 44: 402-9.

[56] Nameda S, Miura NN, Adachi Y *et al.* Antibiotics protect against septic shock in mice administered beta-glucan and indomethacin. Microbiol Immunol 2007; 51: 851-9.

[57] Nameda S, Miura NN, Adachi Y *et al.* Lincomycin protects mice from septic shock in beta-glucan-indomethacin model. Biol Pharm Bull 2007; 30: 2312-6.

[58] Grobe N, Zhang B, Fisinger U *et al.* Mammalian cytochrome P450 enzymes catalyze the phenol-coupling step in endogenous morphine biosynthesis. J Biol Chem 2009; 284: 24425-31.

[59] Shedlofsky SI, Israel BC, Tosheva R *et al.* Endotoxin depresses hepatic cytochrome P450- mediated drug metabolism in women. Br J Clin Pharmacol 1997; 43: 627-32.

[60] Watkins PB. Role of cytochromes P450 in drug metabolism and hepatotoxicity. Semin Liver Dis 1990; 10: 235-50.

[61] Saito M, Nameda S, Miura NN *et al.* Effect of SPG/indomethacin treatment on sepsis, interleukin-6 production, and expression of hepatic cytochrome P450 isoforms in differing strains of mice. J Immunotoxicol 2009; 6: 42-8.

[62] Eum HA, Yeom DH, Lee SM. Role of nitric oxide in the inhibition of liver cytochrome P450 during sepsis. Nitric Oxide 2006; 15: 423-31.

[63] Suda M, Ohno N, Adachi Y *et al.* Tissue distribution of intraperitoneally administered (1-->3)-beta-D-glucan (SSG), a highly branched antitumor glucan, in mice. J Pharmacobiodyn 1992; 15: 417-26.

[64] Nagi N, Ohno N, Tanaka S *et al.* Solubilization of limulus test reactive material(s) from *Candida* cells by murine phagocytes. Chem Pharm Bull (Tokyo) 1992; 40: 1532-6.

[65] Suzuki I, Hashimoto K, Ohno N *et al.* Immunomodulation by orally administered beta-glucan in mice. Int J Immunopharmacol 1989; 11: 761-9.

[66] Sakurai T, Hashimoto K, Suzuki I *et al.* Enhancement of murine alveolar macrophage functions by orally administered beta-glucan. Int J Immunopharmacol 1992; 14: 821-30.

[67] Uchiyama M, Ohno N, Miura NN *et al.* Solubilized cell wall beta-glucan, CSBG, is an epitope of *Candida* immune mice. Biol Pharm Bull 2000; 23: 672-6.

[68] Ishibashi K, Yoshida M, Nakabayashi I *et al.* Role of anti-beta-glucan antibody in host defense against fungi. FEMS Immunol Med Microbiol 2005; 44: 99-109.

[69] Yoshida M, Ishibashi K, Hida S *et al.* Rapid decrease of anti-beta-glucan antibody as an indicator for early diagnosis of carinii pneumonitis and deep mycotic infections following immunosuppressive therapy in antineutrophil cytoplasmic antibody-associated vasculitis. Clin Rheumatol 2009; 28: 565-71.

[70] Harada T, Nagi Miura N, Adachi Y *et al.* Antibody to soluble 1,3/1,6-beta-D-glucan, SCG in sera of naive DBA/2 mice.

Biol Pharm Bull 2003; 26: 1225-8.

[71] Harada T, Miura NN, Adachi Y *et al.* IFN-gamma induction by SCG, β(1-3)-D- glucan from *Sparassis crispa*, in DBA/2 mice in vitro. J Interferon Cytokine Res 2002; 22: 1227-39.

[72] Harada T, Miura N, Adachi Y *et al.* Effect of SCG, β(1-3)-D-glucan from *Sparassis crispa* on the hematopoietic response in cyclophosphamide induced leukopenic mice. Biol Pharm Bull 2002; 25: 931-9.

[73] Tsuji RF, Kikuchi M, Askenase PW. Possible involvement of C5/C5a in the efferent and elicitation phases of contact sensitivity. J Immunol 1996; 156: 4444-50.

Index

A

Adjuvant 39, 42, 44, 48-51, 56, 59, 72, 73
Antibodies 11-13, 21, 39, 50, 72, 76, 77
Autoimmune disease 42, 49, 68, 72, 76

B

Blood sugar 13, 14
Branching 2, 6, 14, 20, 48, 59, 78,
BRMs 10, 39

C

Cancer 1, 3, 10, 13, 21, 27, 39-44, 48, 50, 55, 73, 76,
C3 2, 11, 13, 27, 39-44,
Cell wall 3, 4, 6, 10, 20, 25-28, 39-40, 49, 55-56, 59, 68-69,
Cholesterol 13, 14, 21, 27, 48-49, 55,
Clinical trials 12, 13, 21, 39, 42-44, 49-50
CR3 11, 13, 16, 20-21, 27, 39-44, 49, 58
Cytokine 12, 19-20, 22-23, 25-27, 29, 40, 42, 44, 55, 68-70, 73-75
Curdlan 10, 20, 42, 48, 51, 53, 73

D

Dectin-1 16, 19-26, 29, 49, 55, 58, 69-70
Dendritic cells 21-25, 27-28, 49, 55, 58, 68-69,
DNA delivery 51, 55, 58
Drug delivery 48, 49, 51-61

E

EGFR 41

F

Fish 21

G

Gels 48, 51-53, 61
Genes 16, 22, 59, 71, 77
GM-CSF 22, 69, 70, 77

I

Indomethacin 51
Intestinal bacteria 73
Immunomodulation 3, 26
Infection 1, 2, 6, 12, 17, 19, 21, 25-26, 29, 49, 50, 70, 73-75, 78

L

Lactosylceramide 20, 27, 28, 69
Langerin 20, 28

www.ingramcontent.com/pod-product-compliance
Lightning Source LLC
Chambersburg PA
CBHW041722210326
41598CB00007B/749